Praise for

DREAMS OF IRON AND STEEL

"Smoothly written and informative. . . . Cadbury manages the not-inconsiderable feat of writing lucidly about complex scientific and technological matters, and marches briskly through each of these seven dramatic narratives . . . thrilling."

—*The Washington Post*

"In Ms. Cadbury's recounting of the seven wonders of the nineteenth and twentieth centuries, we find obsessed visionaries engrossed in their own creations who see what others don't and drive their plans forward relentlessly. All her stories involve stupendous costs, and not just economic ones. . . . *Dreams of Iron and Steel* tell[s] the stories behind [these] modern achievements, revealing as much about the human spirit as about technological progress."

—*The Wall Street Journal*

"Deborah Cadbury is a stylish writer and she describes the splendors and miseries of these wonders in their true colors and with understanding of the ethics and aspirations of their times."

—*New Scientist*

"Cadbury has a knack for providing interesting asides . . . an engaging and informative read."

—*San Francisco Chronicle*

"A fascinating look at the technological triumphs of the nineteenth century."

—*Booklist*

"Deborah Cadbury has written an engrossing account of seven great engineering feats of the nineteenth and twentieth centuries. In all but one case, the final result was resounding success; yet Cadbury artfully tells each story so the outcome remains uncertain, in doubt, until the end. A reader is brought directly inside the historical process, experiencing its details just as the contemporary layers did: all the technical mysteries and false steps, the wrenching labors and human costs, and the invincible persistence that finally won the day. Ultimately *Dreams of Iron and Steel* celebrates the triumphs not just of engineering but of the questing human spirit."

—STEPHEN FOX, author of *Transatlantic*

JERRY BAUER

About the Author

DEBORAH CADBURY is a highly praised historian and an Emmy Award–winning producer for the BBC. She is the author of three books, including *The Lost King of France*.

Also by Deborah Cadbury

THE LOST KING OF FRANCE

TERRIBLE LIZARD

THE ESTROGEN EFFECT

SEVEN WONDERS OF THE

NINETEENTH CENTURY, FROM THE

BUILDING OF THE LONDON SEWERS

TO THE PANAMA CANAL

DEBORAH CADBURY

DREAMS of IRON and STEEL

Perennial
An Imprint of HarperCollins*Publishers*

First published as Seven Wonders of the Industrial World
in 2003 in Great Britain by Fourth Estate.

First U.S. edition published in 2004 by Fourth Estate,
an imprint of HarperCollins Publishers.

P.S. ™ is a trademark of HarperCollins Publishers Inc.

HarperCollins books may be purchased for educational, business,
or sales promotional use. For information please write:
Special Markets Department, HarperCollins Publishers Inc.,
10 East 53rd Street, New York, NY 10022.

First Perennial edition published 2005.

Book design by Barbara M. Bachman

Library of Congress Cataloging-in-Publication Data
is available upon request.

ISBN 0-00-716307-X (pbk.)

09 ❖/RRD 10 9 8 7 6 5 4 3

CONTENTS

INTRODUCTION

THE GREAT ACHIEVEMENTS celebrated in this book reveal as much about the human spirit as they do of technological endeavor. The period of over 125 years from the beginning of the nineteenth century saw the creation of some of the world's most remarkable feats of engineering, now celebrated as great wonders of the world, from Isambard Kingdom Brunel's extraordinary *Great Eastern*, the "Crystal Palace of the Seas," which he hoped would join the two ends of the British empire, to the Panama Canal, which linked the Atlantic and Pacific oceans in 1914.

The slowly evolving Industrial Revolution was the fertile ground that gave life to these dreams in iron, cement, stone, and steel. The pioneers of the age were practical visionaries, seeing beyond the immediate horizon, the safe and the known, taking risks and taking society with them as they cut a path to the future. Yet their unique masterpieces could never have been built without an army of unsung heroes, the craftsmen and workers also willing to take risks as they labored to bring each dream to life. And as each great scheme unfolded, the financiers and shareholders were there, too, caught up in the exuberant process and hanging on for the ride as reputations were lost and won.

The Bell Rock Lighthouse—the oldest wonder featured in this book—was created while Britain was in the grip of the Napoleonic Wars. In 1807, Robert Stevenson, the grandfather of Robert Louis Stevenson, who

wrote *Treasure Island*, started work on the Bell Rock Lighthouse in wild
northern seas off the east coast of Scotland. For years he had longed to
make his mark on the world, bringing light to the treacherous Scottish
coast. He dreamed of taking on the most dangerous place of all: the Bell
Rock, a large reef 11 miles out to sea positioned right in the middle of the
approach to the safe haven of the Firth of Forth. Over the centuries, this
deadly reef, submerged at high tide, had cost so many lives it, according
to Robert Louis Stevenson, "breathed abroad an atmosphere of terror"
along the whole coast.

Like so many pioneers of the Industrial Revolution, Stevenson had
cheap labor available, men desperate for work and often prepared to risk
their lives for a meager wage. This was a time when coal and iron ore
were mined by hand and canals were dug with picks and shovels. The
cotton factories, railways, and shipbuilding all needed a plentiful supply
of labor. The population was rising rapidly, in England and Wales alone
from under 7 million in 1750 to 18 million a century later. And because of
recent improvements in agricultural husbandry, rich landowners and
farmers could now produce more with fewer workers. As the rich
enclosed their land, and the old medieval field strips worked by a peasant
population for centuries disappeared, a new landless class laboring for a
wage emerged.

The opportunities of the town beckoned, drawing wave upon wave
of willing recruits. During the nineteenth century the major cities grew,
doubling and redoubling, like cells dividing. Yet at this stage, industrial-
ization had yet to bring real benefits to the workingman. Even after the
1832 Reform Act, a workingman had few rights; he was unlikely to qual-
ify for the vote, and government sympathies lay with his employers. Just
moving to one of England's growing cities could *lower* life expectancy,
which for a laborer was rarely much more than 35 years, and in some
cities, like London or Liverpool, lower still. Many women died giving
birth, and although there were wide variations, records show that in

cities like Manchester, almost 60 percent of children from poor families did not reach 5 years of age, dying mostly from infectious diseases. The poor lived in pitiful squalor difficult to imagine, with several families sometimes sharing a single room. Wages were low; there were no unions, pensions, or Social Security, and no minimum age for labor. Amongst the poorest families everybody was obliged to work: men, women, and children.

Children as young as 9 or 10 were employed in building Isambard Kingdom Brunel's colossal ship, the *Great Eastern*, which was built on the banks of the Thames between 1853 and 1858. They were essential, working in the confined space of the unique double hull of this grand ship, heating and handling thousands of white-hot rivets, their childish voices drowned out by the endless metallic banging and hammering of the iron plates. Horrific accidents were all too frequent, but a death simply meant that there was employment for another child, and there was no shortage of willing workers. There is a noticeable absence of records though—the names and wages of the unsung laborers not even worthy of a note in the record books of the great companies or journals of leading men.

Brunel hoped the *Great Eastern* would be his masterpiece, which would link the ends of the empire. At a time when most ships moored in the Thames were nearer 150 feet in length, built to traditional designs in wood, and powered by sail, Brunel's "Great Ship," was almost 700 feet long, a floating island made of iron, which he envisaged could carry 4,000 passengers in magnificent style as far as the antipodes without needing to refuel. This Leviathan of her day broke the mold of convention; the design was revolutionary, featuring a double hull that made it unsinkable, and powered by enormous engines as high as a house. Every part of the ship's design had been subject to Brunel's penetrating scrutiny, and so complex was the technology that it was claimed only he understood it entirely. He faced enormous criticism: his ship was too big, it was too expensive; it would sink, or break its back on the first big wave—if, that is, he could actually manage to launch it on to the Thames. In fact, it became the blueprint for ship design for years to come.

As the *Great Eastern* became the talk of England, people came to gaze in disbelief at its vast proportions as it miniaturized the working world around it. Even Queen Victoria herself was tempted to venture down the Thames to visit the ship that symbolized the "moral superiority" of her empire. Yet records show that as the queen sailed down the Thames to visit the site, she was obliged "to smell her nosegay *all the time*." For while Brunel was building his masterpiece, the city was in crisis. London was drowning in a sea of excrement.

There were some 200,000 cesspools across the capital, and as the population escalated in the first half of the nineteenth century, so did the smell. In poor districts, these cesspools were seldom emptied, leaving the sewage to overflow, seeping through cracks in floorboards or even running down walls, spreading everywhere with its creeping tentacles of disease. Three epidemics of cholera had swept through London by 1854, leaving over 30,000 dead. The desperation of the poor of the east end even reached the *Times*, in a famous protest: "Sur,—May we beg and beseech your proteckshion and power . . . We live in muck and filthe. We aint got no privies, no dust bins, no drains, no water splies, and no drain or suer in the hole place. . . . The Stenche of a Gully-hole is disgustin. We all of us suffur, and numbers are ill, and if the Colera comes Lord help us . . ."

In the summer of 1858, while the *Great Eastern* was being fitted out for her maiden voyage, the "great stink" finally became unbearable. Joseph Bazalgette, chief engineer of the Metropolitan Board of Works, proposed a grand scheme to build 82 miles of intercepting sewers, a sewage superhighway that linked with over 1,000 miles of street sewers to provide an underground network beneath the city streets. He drove himself to the limits of endurance struggling underneath London's dense housing to create the world's first modern sewage system. The task was made even more difficult since he was in competition with the new underground railway, a network of roads and the emerging overland railway systems. But his ambitious design transformed the city into the first glittering modern metropolis, setting a standard that was quickly copied the world over.

While an endless supply of cheap labor was the human capital for many projects such as the London sewers and the Great Ship, this workforce needed new materials to build for a new age. The iron industry was booming and formed the basis of the Industrial Revolution. During the eighteenth century, output had been modest, with the indigenous ore in England of such low grade that it took nearly seven tons of coal to refine one ton of iron; but by the mid-nineteenth century, improvements, particularly in the use of steam power for blast furnaces, enabled a better quality product to be made more economically. The search for coal and iron ore was ceaseless; by the 1850s, a new blast furnace was opened every two weeks.

The railways were the first of the big adventures in iron. With the invention of the steam engine and the laying of track came untold wealth. As the countryside was opened up and cities were linked, the new roaring engines began shrinking space. Suddenly the country was mobile. Only 500 miles of track had been laid in 1838. Less than 15 years later, by the time of the Great Exhibition in 1851, over 6,000 miles of track crisscrossed the country.

The Golden Age of Railways in Britain was overtaken by the phenomenal growth of railways in America. In the early part of the nineteenth century, the vast continent lay as it had for centuries, marked only by Native American and buffalo trails and the worn wagon tracks of those making the journey west. The eastern and western states of America were still separated by a harsh wilderness that took around six months to cross by wagon. Many preferred to make the journey to California by sea, braving a voyage around South America's Cape Horn rather than risking the dangerous overland crossing. In 1830, when the first American-built locomotive, Tom Thumb, came into service, there were less than 30 miles of working track in the states. Growth was so fast that by 1850 there were over 9,000 miles of track and, by 1860, a staggering 30,600.

The Transcontinental Railway was built with government help during and after the Civil War in the 1860s, and opened up the continent more quickly than a prairie fire, allowing the virgin acres to be settled. President Abraham Lincoln was determined on building a railway line across the continent, "which was imperatively demanded in the interests of the whole country." There were two teams, one building from the east, and the other from California in the west. Of the large numbers who drove the railways across America, the Chinese in the west fared particularly badly, perishing in the thousands in the difficult terrain of the Sierras, their nameless bones gathered and shipped back to China by the crateload. Seven years later, in 1869, the tracks joined, shrinking the whole continent, as the journey from New York to San Francisco could now be done in a matter of days. In a record-breaking run, in 1876, tracks were cleared for the Lightning Train, which raced from coast to coast in just 83 hours. As President Lincoln had envisaged, the Transcontinental Railroad became a catalyst for the vast expansion that would help to make America the industrial giant of the world.

With the growth in cities and improvements in transport, the demand for goods grew. Commerce prospered, trade increased, and more goods were exported. The relentless quest for profit created new wealth and capital, which in turn sought outlets and opportunities for further gain. There was a need to import more raw materials and export the growing surplus of manufactured goods. Wool had traditionally, since the medieval period, been the major trading commodity with Europe, but this had fallen away. In its wake came a demand for more exotic goods to trade in Europe for iron ore and timber. The more adventurous among the merchant traders were landing sugar and cotton, spices and tobacco from the West Indies and America.

With the boom in world trade, by the late nineteenth century, shipping was big business. In Egypt, the Suez Canal had shortened the journey to India, Australia, and the Far East, making trade easier and cheaper, and the world itself a smaller place. Having completed the Suez Canal in 1869, French entrepreneur Vicomte Ferdinand de Lesseps dreamed of an even bolder scheme. He would cut a path across the Isthmus of Panama and unite the great oceans of the Atlantic and Pacific, "from sea to shining sea." The long and dangerous journey around South America and Cape Horn would become a danger of the past. Ferdinand de Lesseps's dream became a symbol of French national pride in the 1880s, and thousands flocked to help with construction. But once out in the tropical heat of Panama, they found themselves facing impenetrable jungle, deep swamps, and deathly tropical diseases as it proved to be an undertaking of nightmare proportions. With 22,000 dead and the investors bankrupted, the canal company failed in 1889 and de Lesseps died a defeated man soon after. Twenty-five years later, the Americans under Colonel George Goethals finally completed the project, opening up new regions for the ever-increasing world trade.

By the 1880s, in spite of the enormous wealth created by the Industrial Revolution in Europe, there were still large numbers in poverty who had not benefited at all. The American economy, however, was growing rapidly, exporting grain and manufactured goods to Europe. And there were plenty in Europe who dreamed of returning on these ships to reach the promised land of America. They had heard of the Statue of Liberty with the words etched on it: "Bring me your poor . . ." They came from Ireland, to escape the potato famine; from England, to look for a better life; and from Russia, to escape the pogroms. Growth in America was so rapid that the population increased tenfold, from 4 million in 1790 to over 40 million by the time of the Centennial Exhibition of 1876.

One such immigrant was John Augustus Roebling, from Muhlhausen, Germany, a brilliant man of enormous determination who won the contract to build the biggest bridge in the world across the wide and turbulent East River, which separates Manhattan from Brooklyn. Roebling's Brooklyn Bridge would be a suspension bridge of great strength and exquisite symmetry: his masterpiece. The foundations alone would reach up to 70 feet below the river; its two mighty towers, at 276 feet, would dwarf much of New York. The total length of almost 6,000 feet seemed a miracle, and all to be built out of a new material: steel.

But this ambitious dream was to cost him the extreme price of life itself; and unknowingly, he condemned his son to a shadow life. Determined to continue with his father's vision, Washington Roebling had to face the horrors of a mysterious new disease, "caisson disease," or as it is now known, the bends. As he and his team labored deep beneath the East River in the hot, humid underground world of the caissons, no one knew who might be struck down next with the terrifying symptoms, paralysis, or even death. Eventually, Washington Roebling succumbed to the mysterious new disease. He was too weak to leave his room and could only continue his work on the bridge by dictating his instructions to his wife, Emily. As the great network of cables was spun across the great East River, he watched through a telescope from his window.

"During all these years of trial and false report," declared one leading official in his praise at the opening ceremony on May 24, 1883, "a great soul lay in the shadow of death, praying only to stay long enough for the completion of the work to which he devoted his life." It had taken 14 years to build the Brooklyn Bridge, but now it transformed the New York landscape and became a triumphant symbol of what men could achieve. As the public fell in love with the sheer American audacity of the enterprise, the heroism cemented into its very fabric, it came to represent "a monument to the moral qualities of the human soul."

With improvements in travel and the growth in prosperity, people found their way across the vast continent. They marked the empty plains with their communities, building the first towns, settling the virgin land, creating a country, stopped only by a poor or hostile environment, such as the desert regions of Arizona and Nevada. And even here, in a region so bereft of life, in the early 1900s Arthur Powell Davis of the U.S. Bureau of Reclamation realized it would be possible to make the desert bloom. He dreamed of harnessing the Colorado River, as it gouged its way for 1,400 miles through snowy heights and forbidding canyons, and turning the wild spirit of the unruly river with its tempestuous floods and mean droughts into an obedient force for good.

The scale of the enterprise was so vast that it took years to win financial backing and government support. Everything about it broke records. As tall as a 60-story building and with a larger volume than the Great Pyramid at Giza, the Hoover Dam, begun in 1931, was the biggest dam in the world. At the height of the Depression, poverty-stricken workers earning just a few dollars a day died from horrific explosions, carbon monoxide poisoning, and heat exhaustion. It was Chief Engineer Frank Crowe, with his skilled management, who built it ahead of schedule and underbudget, and notched up one more extraordinary piece of evidence for the ingenuity of man.

By the time President Franklin D. Roosevelt inaugurated the Hoover Dam in 1935, the last wonder described in this book, the world was transformed in almost every way possible. People's standard of living had increased greatly, average life expectancy had almost doubled in the West, and infant mortality had virtually disappeared. Other systems of transport had been developed, too, including the automobile, which gave many people their own private transport. Higher education and specialist training opened up new opportunities for those whose forebears a few generations before had been laboring in fields unable to read or write. The £1 a week that Robert Stevenson had given his laborers to work a 12-hour day, seven days a week, wet or dry, had, by the time the Hoover

Dam was lighting up the western deserts, turned into a wage that a work-ingman, increasingly backed by unions, could live on more comfortably.

The stories in this book capture the restlessness and ambition of an age and represent a high watermark of industrial achievement. Each wonder illustrates the swiftly moving frontiers of technology and serves as a unique monument, a marker for what was known at the time. Taken together, the wonders illustrate progress by charting the frontiers of industrial knowledge and expertise. Timing is critical; it is no accident that these particular achievements occurred when they did. It would not have been possible to create a Hoover Dam or a Panama Canal earlier in the nineteenth century. Indeed, the French tried and failed in Panama, since the technology and infrastructure to create the canal was not in place. The changes are not linear; history is not about a smooth, even progression. There are enormous bursts of creative endeavor and change that reach out in unexpected directions until what was once barely possi-ble becomes routine.

In one sense, the stories told here present a romantic view of man—of an individual who struggles to realize his dream and make a mark on the world. As the nineteenth century progressed, the men of genius took the stage in quick succession, each engrossed in his own creation to the exclusion of all else. Each in turn gave so much of himself, often denying relationships, sleep, basic human comforts, and ultimately in some cases, his health; all thought and energy given over to the demands of his creative work. Robert Stevenson struggled in wild seas to create his lighthouse, which is the oldest offshore stone lighthouse still stand-ing anywhere in the world, a testimony to his battle with the sea. The Roeblings—father and son—were prepared to give their very lives to the Brooklyn Bridge, whose exquisite symmetry and beauty has since inspired poets and artists. And Brunel, in the very midst of the Industrial Revolution, seemingly directing it himself, throwing his small, energetic figure into the great mélange, absorbed in the delight of it all, unable to tear himself away from his Great Ship, no matter what was the price to be paid.

The legacy of their great ambition and talent remains to this day. With the exception of Brunel's Great Ship, all the wonders have survived to the twenty-first century and are now celebrated as powerful symbols of the modern world. The wealth of inspiration and energy of the nineteenth century was the catalyst for the huge progress that marked the twentieth century, as the coming industrial giants stood on the shoulders of an earlier generation.

CHAPTER 1

THE GREAT EASTERN

I have never embarked on any one thing to which
I have so entirely devoted myself, and to which I have devoted
so much time, thought and labour, on the success of
which I have staked so much reputation . . .

ISAMBARD KINGDOM BRUNEL
on the Great Eastern, *1854*

IN 1857, ISAMBARD KINGDOM BRUNEL, Britain's foremost engineer, paused one day for a photograph in Napier's shipyard at Millwall in East London. Cigar in mouth, with mud caked on his shoes and trousers, this is no formal photograph. He has his hands in his pockets, his clothes are creased, his hair untidy. The face and, particularly, the eyes are absorbed in something that can only be imagined, something that occupies him completely. He looks like a man with a future.

Brunel was at the peak of his fame, his latest venture had become the talk of England. Behind him rose the massive dark shape of the hull of the *Great Eastern*, the largest ship in the world, the Leviathan of her day. Expectant sightseers from across Europe came to see her on the banks of the Thames, where she rose, wrote Charles Dickens, "above the house-tops, above the tree-tops, standing in impressive calmness like some huge cathedral." Nothing like this had been seen before; when complete, she would be the largest moving man-made object ever built and, for many, a symbol of the greatness of the British Empire. Yet far from being the final triumph in Brunel's brilliant career, the *Great Eastern* was to become the monstrous creation that would destroy him.

Brunel's grand scheme had begun to take shape a few years earlier, shortly after the Great Exhibition of 1851. Britain had seen a spectacular boom in the railway industry, with over 6,000 miles of track laid since the 1830s. Brunel himself, caught up in the thick of railway mania, was increasingly disillusioned by it. "The whole world is railway mad," he protested to a friend. "I am really sick of hearing proposals made." Amongst his sketches for Paddington station in London in 1852, his notebooks reveal drawings of his next bold venture: a great steamship, almost twice the length of any ship ever built.

He dreamed of a floating city, majestic by day and a brilliant mirage at night, reflecting a million lights in the dark water. It was to be a ship of such vast and unheard of proportions that she would be able to carry 4,000 passengers in pampered luxury as she steamed through distant seas. In the evening there would be dancing under sparkling chandeliers or a stroll on deck in especially manufactured "moonlight" as she pursued her steady course to the antipodes magically, without need of refueling. But could Brunel ever realize his dream and build the "Crystal Palace of the Sea"? Who could afford to support him?

Most ships docking in the Thames in the mid-nineteenth century were made of wood and built to a traditional design around a skeleton of wooden ribs giving strength to the hull. They were wind-powered and usually little more than 150 feet in length. Brunel's "Great Ship," as she came to be known, was to be 692 feet long, 120 feet wide, and 58 feet deep. Enormous engines as high as a house, with the power of over 8,000 galloping horses, would drive her paddle wheels and screw engines. In addition, an impressive 6,500 yards of sail would be carried on six masts and five funnels that were spread along her deck. The revolutionary new design of her hull, strong and streamlined, would see her cut through the seas as smoothly as a knife cuts butter, and she would have the practical capability to carry all her own fuel to the farthest reaches of the empire and back. Brunel felt certain there was a need for such a ship.

To most shipbuilders of the day, Brunel's Great Ship would have seemed an unattainable, magnificent dream, but Brunel had a way with dreams—his *châteaux d'Espagne*. At 20, he had risked his life as engineer on the first tunnel under the Thames. At 24, he was elected a member of the Royal Society. He went on to design five suspension bridges, including the magnificent Clifton Suspension Bridge at Bristol, as well as wet and dry docks, tunnels, and piers. In 1833, he became engineer-in-chief to the Great Western Railway. His trains, speeding at 50 miles an hour past fields and villages, opened up England and joined distant cities. He surveyed and planned the route from Paddington to Bristol, designing the track, cuttings, tunnels, stations, and trains—even the signal boxes and at least 125 bridges on the route west. As he built the Great Western Railway, his dreams became even grander. He imagined large steamships that would go farther west, sailing out across the Atlantic to America, shrinking the world's oceans and creating a global system of transport.

Brunel launched the *Great Western* in 1837, which, although constructed traditionally in wood, was the largest steamship ever built, and faster, too. In 1843, he surpassed this with his second ship, the *Great Britain*, which at nearly 322 feet long and 50 feet wide, was the first large iron screw steamship, with a new design of hull and increased emphasis on longitudinal strength. The *Great Britain* was not built in the traditional rib design; instead, 10 iron girders ran the length of the vessel at the base. Everything about the ship was designed for strength, and she became an extremely profitable ship on the Australia run. In 1851, with the discovery of gold in Australia and increasing demand for passenger and mail traffic to the east, Brunel could see a commercial possibility for an even bigger ship. "Size in a ship is an element of speed," he argued, "and of strength and of safety and of great relative economy."

In the spring of 1852, Brunel approached the leading naval architect of the day, John Scott Russell, to seek advice about the feasibility of his plan for the Great Ship. Scott Russell had made his name streamlining the design of ship hulls for maximum efficiency with his "wave-line theory." Brunel admired him and respected his achievements, notably his original work on hull design. "With respect to the form and construction of the vessel itself," said Brunel, "no body can, in my opinion, bring more scientific and practical knowledge to bear than Mr. Scott Russell."

Scott Russell was equally impressed with Brunel and his extraordinary idea for the Great Ship. He suggested that they present the plan for the vessel to the Eastern Steam Navigation Company. This company had recently lost the mail contract to the Far East to Peninsular and Oriental Steam Navigation Company and were thought to be looking for a way to revive their fortunes on the route to India, China, and Australia. Although a committee at the Eastern Steam Navigation Company decided in favor of the plan and appointed Brunel as engineer, the proposal caused a heated debate, with some directors resigning. Indeed, Brunel soon learned that "certain of the directors . . . opposed me personally."

Undeterred, he threw himself into finding others interested in the project, and by the time capital for the venture was found, more than half of the new directors had been nominated by Brunel. Charles Geach, M.P. for Coventry and chairman of the Midland Bank, was an important member of the committee and a business associate of Scott Russell, to whom he supplied iron from a smelting company where he was a partner. Brunel, aware that most of the directors were of his choosing, felt "the heavy responsibility of having induced more than half the present directors of the company to join." The fact is, he admitted, "I never embarked on any one thing to which I have so entirely devoted myself and to which I have devoted so much time, thought and labour, on the success of which I have staked so much reputation."

By May 1853, enough shares had been sold to make a start, and Scott

Russell presented his estimate to the board for £377,200 to build the ship. The hull was estimated to cost £275,200, the screw engines £60,000, and the paddle engines and boilers £42,000. Since Brunel had calculated the cost of the ship as somewhere near half a million, the board was delighted with Scott Russell's price. Scott Russell had, however, seriously underestimated the costs. He wanted the commission so much that it is possible that he put in a low bid to ensure he got it, but the subsequent shortage of funds was to put the venture in jeopardy.

Soon Scott Russell faced further difficulties when, in September 1853, fire broke out in his shipyard. As the flames licked hungrily around the carpenters' shop, the boilermakers' shop, and the timber yard, the London fire brigade "hurried towards the seat of the calamity . . . [but] unfortunately the light of the fire deceived them," reported the *Times*. They galloped at full speed down the wrong side of the river and were almost at Deptford before they realized where the fire was raging. In their rush back across London Bridge, "a body of the fire brigade were nearly killed as the horses galloped with the engine into a hole eight feet deep, pitching all the firemen off the machine." It took over an hour for the fire engines to reach the scene, by which time the fire was well entrenched. By the light of day, everything in Scott Russell's yard was seen reduced to charred, black, smoking stumps. The damage was estimated at around £140,000, and most of the yard had not been insured.

A worrying and faintly discordant background to the ambitious project was becoming discernible. Scott Russell's yard was temporarily demolished, and he was financially embarrassed. His yard lacked supervision, too, and with no provision for safeguarding large supplies, pilfering was rife. And despite the fact that the *Great Eastern* was twice the size of any previous ship, she was to be built in a traditional shipyard with no special arrangements made for her construction. Against this vaguely uneasy situation, Brunel insisted on total control.

Finally, in December 1853, all parties signed contracts for the Great

Ship. Scott Russell was to build the hull and paddle engines, the celebrated James Watt and Co. would build the screw engines, and Brunel was to approve all changes and drawings at every stage of the construction process. "I cannot act under any supervision," he said, "or form part of a system that recognises any other advisor than myself." Indeed, everything concerning the Great Ship was to be "to the entire satisfaction of the engineer."

However, conflicts soon arose over the methods of payment. Brunel wanted to pay Scott Russell as work progressed, calculated on the amount of iron put into the ship on a monthly basis; this was the custom adopted for his railway contracts. Scott Russell, meanwhile, expected to be paid lump sums on a more regular basis so that he had sufficient cash to meet expenses. Liquidity, which had always been a problem for him, was now made worse, as some of his payment for work on the hull was in company shares, although Charles Geach helped by taking shares as payment for his iron instead of cash.

The most immediate practical concern for Brunel was where to build and launch his ship. The contract had stipulated that the ship be built in a dry dock, but Scott Russell's yard was too small for Brunel's vast project. Scott Russell, needing more space, rented the yard adjoining his, Napier Yard, at Millwall, to accommodate the construction of the hull. They realized that building the ship on an end slip was impossible; the bow would be 100 feet in the air during construction. It was also out of the question to launch a 700-foot-long ship stern first in the traditional way, when at that point, the river was only 1,000 feet wide. After much deliberation, Brunel decided to build the ship on the bank parallel to the river and launch her into the Thames sideways down a gentle slope. To prepare the site, 2,000 oak beams up to 40 feet long were piled 5 feet apart into the shore, leaving 4 feet aboveground. Further packing was then added, and the flat-bottomed hull would rest on two large cradles.

In March 1854, just as they were finally ready to start construction work on the hull, Britain entered the Crimean War. With supplies needed for the military, the price of iron rose rapidly. A few months later, Scott Russell faced another major setback: the unexpected death of Charles Geach. "The Honourable Member had returned from his Scotch shooting in unusually good health," reported the *Times*. However, he had been suffering from a dangerous infection in his leg. A change for the worse occurred, "and after much suffering borne with the greatest fortitude and resignation, Mr. Geach expired at half past 4 o'clock yesterday afternoon."

This was a disaster for Scott Russell. Geach was crucial to the iron supply and was the one person who had accepted flexible terms of payment on this immensely expensive project. Scott Russell's solution was a dangerous one. Unknown to Brunel and the other directors, he secretly mortgaged his shipyard, thereby putting the Great Ship at risk. Brunel had not worked closely with Scott Russell on a financial venture before and was unaware, at this point, of the danger signals beginning to emerge that could affect the future of his "Great Babe," as he affectionately nicknamed his new creation.

IN SPITE OF THE precarious financial situation, the hull was slowly rising, a massive dark shape against the water and sky. It was being built at a big southerly bend in the river at Millwall on the Isle of Dogs, about six miles downriver from Westminster and 40 miles to the open sea. This was an undeveloped part of London, previously the haunt of wildlife, with, as local accounts put it, little but "marshy fields and muddy ditches, with here and there, a meditative cow cropping herbage." Now it was the site of a technological revolution that would inspire shipbuilding for years to come.

Brunel's original idea for the design of the hull was evolved from lessons learned in his bridge-building days. A key innovation was to have a *double* hull, heavily braced up to the waterline, one hull inside the other and 2 feet 10 inches apart. These were to be constructed from 30,000 iron plates, each three-fourths of an inch thick. The deck, too, would consist of two thicknesses of half-inch iron plate. Strength was further guaranteed by longitudinal and transverse bulkheads. According to Scott Russell, "the longitudinal system is carried throughout unbroken, without interruption by the bulk heads"; and the watertight transverse bulkheads, 60 feet apart the length of the ship, made her, it was hoped, virtually unsinkable. With this cellular double hull, the bulkheads, and the strong watertight deck, Brunel compared the strength of the hull to that of a box girder.

This unique vessel was to be powered by both screw and paddle engines to give greater flexibility and maneuverability. The engines themselves would be giants, 40 feet high; the cylinders on the screw engine alone had a bore of 7 feet by 14. And before the crankshaft for the paddle engines could even be made, new larger furnaces had to be built.

The giant hull was attended by an army more than 1,000 strong of riveters, bashers, and shipwrights. Men were busy inside, outside, high up, low down, creeping and crawling between the hulls, up at the bow, down at the stern, hammering, clanking, banging, carting iron, moving wood, and hammering the rivets. Swarming all over the ship, they gave the impression of ants on a giant carcass. Depending on the light, men worked 12-hour days, and a skilled man could earn 30 to 40 shillings a week. But there was no certainty of continued employment, and plenty of men were waiting to step into the shoes of any who left their jobs.

The noise coming from the shipyard was deafening. The ringing sound of metal hitting metal reverberated throughout the hull. Two hundred riveting teams working both inside and outside the hull hammered unceasingly at the 1-inch-thick, white-hot rivets; 3 million rivets in total. Each rivet would be held in place by a man on the other side of the plate. Children were employed as part of the team tending the forge and placing the heated rivet in the hole. They were particularly useful working in the double hull, where limited space made it difficult for an adult man to maneuver.

Working in the dark, confined space of the double hull, it did not do to lose concentration, even after a 12-hour shift: one moment of carelessness could be paid for with a hand, or an arm, or a life. Accidents were commonplace. It was all too easy to miss a step and, falling from a height, involve another man in disaster in the overcrowded conditions. One worker, making bolts, got his hands tangled in the machinery and torn completely from their sockets at the wrist. In his case, amputation of both arms was the only solution. Another man, curious about the working of a pile driver, was bent over examining the machinery when the hammer came down, flattening his head. Children were particularly vulnerable. They could be working in the yard as young as 8 or 10. One unfortunate child fell from a height and was impaled on an upright iron bar. According to one witness, "after he was dead, his body quivered for some time." There was always another boy willing to take his place for a shilling or two. A rumor persisted at this time that a riveter and his boy had somehow been forgotten and were entombed alive into a section of hull. Months later, workers said they could hear the ghosts hammering, trying to escape. Most believed this tragedy would put a curse on the ship.

As work progressed, it became apparent that Brunel and Scott Russell, the two great men locked together in building this ship, were very different in style and temperament. Brunel was married to his work, always absorbed in every detail. He thought nothing of getting dirty in the course of a working day. His wife, Mary, known as "the Duchess of Kensington" on account of her beauty and style, rarely saw him at their elegant Westminster home, as he worked 18 hours a day. Scott Russell, on the other hand, had a more relaxed managerial style; he delegated. As he sat in his impressive office, he expected the chain of command to work perfectly around him. He did not intend to get his beautifully tailored clothes dirty and left the management of the site to his yard managers, Hepworth and Dixon, who were responsible for the shipwrights. Increasingly, Brunel and Scott Russell found themselves in disagreement.

As the work on the Great Ship progressed and her shape became more evident, the press began to take an interest. They estimated that the *Leviathan*, or *Great Eastern*, as she became known, had greater dimensions even than Noah's Ark. So-called *Great Eastern* fever began to spread throughout the country and the ship soon came to symbolize the "moral supremacy" of the British Empire. Yet for all the growing national excitement, a serious rift between Brunel and Scott Russell began to emerge.

An article in the *Observer* in November 1854 sparked the first open clash. The paper had mistakenly credited Scott Russell with playing the major role in the design of the Great Ship. They quoted him "as carrying out the design" and claimed that Mr. Brunel had merely "approved of the project." Brunel was furious, and he wrote to Company Secretary John Yates, correcting this error in no uncertain terms. He strongly suspected that Scott Russell or one of his men had leaked the article. "This bears rather evidently a stamp of authority, or at least it professes to give an account of detail which could only be obtained from ourselves," he wrote. "I cannot allow it to be stated, apparently on authority, while I have the whole heavy responsibility of success resting on my shoulders, that I am the mere passive approver of the project of another, which in fact originated solely with me and has been worked out by me at great cost of labour and thought devoted to it for not less than three years." Quite how the information had reached the *Observer* was never ascertained.

Difficulties increased during the winter of 1854, with Scott Russell facing growing financial problems following the death of Charles Geach. His bank refused to give him any more credit, so he asked the board of Eastern Steam if he could be paid in the future on a regular monthly basis for work accomplished. In April 1855, there was another fire at his yard, and while no damage was done to the Great Ship, he bore a further loss of £45,000. Faced with these difficulties and endless delays in construction at Scott Russell's yard, Brunel was reluctantly forced to concede that the launch date, originally planned for October 1855, would have to be deferred.

A fundamental conflict between the two men arose over the method of launching the Great Ship, too. At an estimated 12,000 tons, this was the largest weight ever moved by man and it had to be moved 200 feet into the river. Brunel had given much thought to the problem and had come to the conclusion that the only way to achieve this was through a "controlled" launch. This was a most unusual procedure. Pushing the mighty 700-foot-long ship sideways into the river seemed fraught with problems; but Brunel, undaunted, insisted it was the only way to launch her. Anything else would be a disaster. He envisaged the possibility of the ship getting stuck or, worse, moving into the river far too quickly and keeling over or breaking up. He preferred to err on the side of caution and planned to use restraining chains to control the ship's progress gently down the slope.

Scott Russell was totally opposed to a controlled launch. He pointed out that "free" launches, admittedly of smaller ships, were carried out successfully on the Great Lakes of America. He was also worried about the cost of a controlled launch. Since Scott Russell was under contract to launch the ship, and Brunel's plan was estimated to cost an extra £10,000, he was not to be moved on this subject.

Through the spring and summer of 1855, Brunel was concerned as Scott Russell became increasingly uncooperative. The latter would be unavailable, slow in replying to letters, and vague with the information that he did give. Brunel needed specific facts that would enable him to work out launching requirements, so his irritation with Scott Russell grew. "I begin to be quite alarmed at the state of your contract," he wrote. "Four months are gone and I cannot say even the designs are completed . . . to justify a single bit of work being proceeded with." In May 1855, he wrote again to Scott Russell in an exasperated tone: "Your reply this morning to my long list of complaints is an admirable specimen of an Under-Secretary's reply in the House to a Member's motion—it does not satisfy one single honest craving for information and for assurance of remedy . . . I do not want to better indicators than usual . . . Those made on this occasion and to which I object were absurd—like the attempts at writing of a two-year-old baby."

As the summer wore on, Scott Russell, faced with yet another fire at his yard, became more immersed in his financial uncertainties. And Brunel was totally consumed with giving life to his creation, transforming so many lifeless tons of iron and wood into the majestic shape of his inner vision. To this end he was always occupied, dealing with endless problems and finding endless solutions. He went to Haverfordwest in Wales to organize jetties where the Great Ship would take on coal. He found the man who he felt had the necessary qualities and experience to captain his great ship: William Harrison. There were also detailed discussions to be had on the design of the engines. Brunel was soon worried to hear that Scott Russell was not fulfilling his contract. It had become apparent that the work on the hull was not commensurate with the amount of money Scott Russell had received. Scott Russell had in fact been paid the bulk of the money, but there was still a massive amount of work to be done before the hull was anywhere near complete. Brunel slept only four hours a night and worked like a man possessed.

By late summer, Brunel was still trying to get information from Scott Russell that might affect plans for the launch of the ship. Again he wrote, requesting information from Scott Russell concerning the center of gravity for the ship, and again he felt the reply he received was too vague. Scott Russell meanwhile wrote with a request for more money, £37,673 to be precise. He argued that this sum was for extra work—alterations that

Brunel had made to the original designs. This led to lengthy arguments and exacerbated the growing distrust between the two men. Scott Russell did finally provide a launch date, for March 1856, but infuriated Brunel by carelessly giving the wrong information on the tonnage of the vessel. Brunel was angry. "How the devil can you say you satisfied yourself at the weight of the ship," he wrote to Scott Russell in October 1855, "when the figures your clerk gave you are 1,000 tons less than I make it or than you made it a few months ago—*for shame*—if you are satisfied. I am sorry to give you more trouble but I think you will thank me for it—I wish you *were* my obedient servant, I should begin by a little flogging."

By now, very little charm was being wasted in dealings between the two men. Scott Russell replied with a request for more money needed to pay the banker, Martins & Co., which held his yard in mortgage. The Great Ship, it seemed, was eating money and he could not obtain credit from anywhere. He wrote again to Brunel insisting on regular payments, saying, "I fear I shall get into trouble unless we can see our way to a *definite* arrangement for the future. I am keeping an enormous establishment of people night and day. I either must have payment with certainty or reduce my number of hands." Trying to ease the strain on his finances, Scott Russell had taken orders for six other smaller ships, which he was building in the yard of the Great Ship, and on which the labor force was increasingly deployed. To add to Brunel's disgust, the smaller craft in the yard were so placed that essential work on the Great Ship was made impossible. And still Scott Russell had not produced the information needed for the launch. Brunel wrote again on December 2, 1855: "I must beg you to let me have with the least possible delay the correct position of the centre of flotation at the 15-foot draft line . . . I cannot stand any longer the anxiety I have felt ever since we commenced the ship as to her launching."

Still no information was forthcoming from Scott Russell. Whether he was deliberately holding back the information Brunel required in the hope that he would be forced eventually into a cheaper uncontrolled launch is not known. But, ultimately, Brunel defeated him; he managed to ascertain the center of flotation and soon had plans prepared for the launching cradles and launchways. Scott Russell complained to the board, pointing out that the controlled launch had not been part of the original plan and that the additional cost he faced was prohibitive.

By January 1856, Brunel had finally had enough. He voiced his concerns to the board of the Eastern Steam Navigation Company, reporting that a "large deficiency" in iron "appeared to exist" at the yard, which was difficult to understand, as "I do not now believe that any mistake has been intentionally made or even intentionally overlooked . . . and have been assured . . . that none of the iron so imported was ever knowingly used for other purposes. . . . I make the quantity in the yard about 1,400 tons but this would still leave 800 or 900 tons to be accounted for and I am totally at a loss to suggest even a probable explanation." To make his position absolutely clear, he went on to say, "I have great cause to complain of neglect or to say the least of it of attention to my orders and remonstrances. My instructions even when repeated frequently and formally in writing are much disregarded. . . . Mr. Russell, I regret to say no longer appears to attend either to my friendly representations and entreaties or to my own formal demands and my duty to the company compels me to state that I see no means of obtaining proper attention to the terms of the contract otherwise than by refusing to recommend the advance of any more money."

Dark clouds were gathering over the shipyard. If Scott Russell was eventually made bankrupt, the Great Ship might belong to the creditors. A crisis was reached when Martins, Scott Russell's bankers, refused to honor his checks. Scott Russell laid off the workers in the yard, saying he could no longer continue with the work. The board of Eastern Steam, on

Brunel's advice, seized the Great Ship, stating that Scott Russell had breached his contract. The creditors were informed, the accountants moved in, and work on the Great Ship came to a complete stop.

The board submitted a claim on Scott Russell's estate, citing breach of contract, only to find there was no estate left to which they could lay claim. Brunel had been unaware that Scott Russell had mortgaged his yard and that there were a number of creditors, Beale and Co., the iron manufacturer, being the largest. Apart from the mounting debt and the missing iron, it emerged that although only about a quarter of the work had been completed on the hull, Scott Russell had somehow been paid £292,295. The board of Eastern Steam and its worried shareholders now found themselves in the hands of Martins's bank, which had prior claims. After much difficult discussion, the bank acknowledged that Eastern Steam had rights too, and agreed to renew the lease of the shipyard—but just until August 1857.

Brunel was understandably worried. "I feel a much heavier responsibility now thrown upon me than I ever intended to take upon myself," he wrote. He still had Hepworth and Dickson from Scott Russell's establishment to work with; and better still, Daniel Gooch, a colleague and old friend involved in his railway commissions, was approved as his assistant. Scott Russell, however, was humiliated. "I intend to be very cautious and to keep every string which it devolves on me to pull, tightly in my own hands," Brunel told him. "It would therefore be in the position of an assistant of mine—that I should propose to engage your services."

The last two years had been immensely difficult for Brunel as he tried to bring his work of creative genius to fruition. The burden of organizing such a vast project against such a fraught background was beginning to exact a toll. He had taken liberties all his life with his strong constitution and robust health, ignoring warnings, winning glory, and generally taking life at a gallop; but now the rumor spread that he was ill. With little over a year left in which to launch his Great Ship before the bank took control of the yard, Brunel was facing the supreme test of his entire career.

"WHERE IS MAN TO GO for a new sight?" asked the *Times* in April 1857. "We think we can say. In the midst of that dreary region known as Mill-wall, where the atmosphere is tarry, and everything seems slimy and amphibious, where it is hard to say whether the land has been rescued from the water or the water encroached upon the land . . . a gigantic scheme is in progress, which if not an entire novelty, is at as near an approach to it as this generation is ever likely to witness." The excitement was tangible; with Brunel in complete charge, work progressed so well that, by June, the ship was almost ready for launching. The *Great Eastern* had become the talk of Europe.

However, the optimism of the summer dissolved as endless difficulties connected with the launch arose. After much negotiation, a new launch date was agreed on with Martins, for October 1857. If the ship was not in the Thames by this date, the creditors would claim the yard and "we will be in the hands of the Philistines," declared Company Secretary John Yates. October 5 arrived, and Brunel, not satisfied that everything was ready for the launch, had no alternative but to defer the date once again. The mortgagees seized the yard and refused access to all working the ship. The situation had become impossible. Brunel was now put under immense pressure to agree to launch on the next spring tide of November 3, and the company was charged large fees for the delay.

As the Great Ship stood helplessly inert, waiting at the top of two launchways, her brooding shape invited much comment. Many thought she was unlaunchable and would rust where she was. Other, wise "old salts" predicted that if she ever did finally find the sea, the first wave would break her long back in half. Brunel never doubted; but no matter how carefully he planned the coming operation, there were still many unknowns and little time to test the equipment. In the small hours of the cold winter nights it seemed he was attempting the impossible. He was proposing to move an unwieldy metal mountain more than four stories high down a precarious slope towards a high tide with untried equipment.

The plan for launching the ship sounded simple. Hydraulic rams would gently persuade her down the launchway. Tugs in the river, under the command of Captain Harrison, could also ease her towards the river, and there were restraining chains to hold her back should she move too fast. Two wooden cradles, 120 feet wide, were supporting the ship and they rested on launchways of the same width. Iron rails were fixed to the launchways and iron bars 1-inch thick were attached to the base of the cradles, and both surfaces were greased to enable the vessel to slide easily down the launchway gradient.

As the day for the spring tide of November 3 approached, work on the ship became more frenzied. At night, 1,500 men working by gaslight carried out last-minute instructions from Brunel. Brunel himself never left the yard, sleeping for a few hours when exhausted on a makeshift bed in a small wooden office. He issued special instructions to everyone involved with the launch, saying: "The success of the operation will depend entirely upon the perfect regularity and absence of all haste or confusion in each stage of the proceedings and in every department, and to attain this nothing is more essential than *perfect silence*. I would earnestly request, therefore, that the most positive orders be given to the men not to speak a word, and that every endeavour should be made to prevent a sound being heard, except the simple orders quietly and deliberately given by those few who will direct."

Unfortunately for Brunel, perfect silence was not a high priority with the board of the Eastern Steam Navigation Company. For months now the company had borne the disastrous hemorrhage of enormous amounts of money to give life to the Great Ship. The launch presented them with a small chance to recoup some of those losses. Unknown to Brunel, they had sold more than 3,000 tickets to view the launch from Napier yard. Newspapers, too, had played their part, informing the public that an event worthy of comparison with the Coliseum was about to take place at Millwall. "Men and women of all classes were joined together in one amicable pilgrimage to the East," reported the *Times*. "For on that day at some hour unknown, the *Leviathan* was to be launched at Millwall . . . For two years, London—and we may add the people of England—had been kept in expectation of the advent of this gigantic experiment, and their excitement and determination to be present at any cost are not to be wondered at when we consider what a splendid chance presented itself of a fearful catastrophe."

The launch place of the *Leviathan* presented a chaotic picture to Charles Dickens. "I am in an empire of mud. . . . I am surrounded by muddy navigators, muddy engineers, muddy policemen, muddy clerks of works, muddy, reckless ladies, muddy directors, muddy secretaries and I become muddy myself." He noted that "a general spirit of reckless daring" seems to animate the "one hundred thousand souls" crammed in and around the yard, upon the river, and the opposite bank. "They delight in insecure platforms, they crowd on small, frail housetops, they come up in little cockleboats, almost under the bows of the Great Ship . . . many in that dense floating mass on the river and the opposite shore would not be sorry to experience the excitement of a great disaster."

In the dull light of the November morning the scene that greeted Brunel as he emerged from his makeshift quarters was one of confusion and noise, with uncontrollable crowds swarming over his carefully placed launch equipment. All of fashionable London, displaying intense curiosity and expecting to be amused, charmed, and hopefully thrilled, was taking the air in Napier yard. Then, almost farcically, in the midst of preparations, a string of unexpected distinguished visitors turned up in all their finery: first the Comte de Paris and, then, complete with a retinue resplendent in gold cloth, the Ambassador of Siam. A halfhearted attempt at a launching ceremony saw the daughter of Hope, chairman of the board, offering the token bottle of champagne to the ship. Brunel refused to associate himself with it. She got the name wrong, christening the ship *Leviathan*, which nobody liked since all of London had already decided on the *Great Eastern*.

The whole colorful fun fair scene was terribly at odds with the cold, clinically precise needs of the launching operation. Brunel felt betrayed; as he later told a friend: "I learnt to my horror that all the world was invited to 'The Launch,' and that I was committed to it *coute que coute*. It was not right, it was cruel; and nothing but a sense of the necessity of calming all feelings that could disturb my mind enabled me to bear it."

Brunel had no alternative but to make the attempt in spite of the difficulties. He stood high up on a wooden structure, against the hull, his slight figure wearing a worn air, stovepipe hat at an angle, the habitual cigar in his mouth. He held a white flag in hand, poised like a conductor, waiting to begin the vast unknown music.

As the flag came down, the wedges were removed, the checking drum cables eased, and the winches on the barges mid-river took the strain. For what seemed an eternity, nothing happened. The crowd, which had been quiet, grew restive. Brunel decided to apply the power of the hydraulic rams. Suddenly, with thunderous reverberation, the bow cradle moved three feet before the team applied the break lever on the forward checking drum. Immediately, accompanied by a rumbling noise, the stern of the great hull moved four feet. An excited cry went up from the crowd: "She moves! She moves!" In an instant the massive chains of the aft checking drum cable were pulled tight, causing the winch handle to spin. As the winch handle "flew round like lightning," it sliced into flesh and bone, and tossed the men who worked the drum into the air like flotsam. The price the team paid for not being entirely awake to the quickly changing situation was four men mutilated. Another man, the elderly John Donovan, sustained such fearful injuries he was considered a hopeless case and was taken to a nearby hospital, where he soon died.

Brunel tried to move the hull once more later that afternoon, but a string of minor accidents and the growing dark persuaded him to give up for the day. In the words of Brunel's 17-year-old son Henry, "The whole yard was thrown into confusion by a struggling mob, and there was nothing to be done but to see that the ship was properly secured and wait till the following morning."

The high tide had come and gone; the next one was not for another month, another month of extortionate fees while the ship lay on the slipway. Brunel was determined to get the ship completely onto the launchway as soon as possible. He was very concerned that while the ship was half on the building slip (whose foundation was completely firm) and half on the launchway (which had more give) the bottom of the hull could be forced into a different shape. His urgent task was to get the ship well down the launchway to the water's edge in case the hull started to sink into the Thames mud under its immense weight. The fiasco of November 3 had, at least, provided information on how to manage the launch more effectively. Some alterations were made to improve the equipment and another attempt at the launch was made on the nineteenth. This was a huge disappointment, with the hull moving just one inch. Clearly more would have to be done.

The winches on the barges mid-river had been ineffective, and all four were now mounted in the yard, their cables drawn under the hull and across to the barges; but even chains of great strength and size broke when any strain was put on them. "Dense fog made it almost impossible to work on the river," observed Henry Brunel. "Moreover, there seemed a fatality about every attempt to get a regular trial of any part of the tackle." Two more hydraulic presses were added to the original two, giving a force of 800 tons at full power.

By November 28, Brunel felt confident enough to try and move the

Great Ship once more. From a central position in the yard, he signaled his instructions with a white flag. The hydraulic presses were brought up to full power and, to the accompaniment of terrifying sounds of cracking timber and the groaning and screeching of metal, slowly the ship moved at a rate of one inch per minute. As before, though, the tackle between the hull and the barges proved unreliable and Captain Harrison and his team found themselves endlessly repairing chains. In spite of the difficulties, by the end of the day, the ship had been moved 14 feet, and there was renewed hope of floating the ship on the high tide of December 2. An early start on November 29 saw the river tackle yet again let them down, and the four hydraulic presses pushed to their maximum could not move the ship. Hydraulic jacks and screw jacks were begged and borrowed, and by nightfall another eight feet had been claimed so that by the thirtieth the ship had moved 33 feet in total and hope now had real meaning. But then one of the presses burst a cylinder, which killed off any possibility of a December launch.

But Brunel would not be defeated, and throughout December he carried on, inching the colossal black ship down the launchway. Each day it was becoming more reluctant to start. "While the ship was in motion," noted Henry Brunel, "the whole of the ground forming the yard would perceptibly shake, or rather sway, on the discharge of power, stored up in the presses and their abutments." In the freezing fogs of December and January, the hundreds of workers were heard rather than seen as the yard echoed with the sounds of orders and endless hammering. The gangs sang to relieve their boredom as they mended the chains. At night, gas flares lit the scene and fires burned by the pumps and presses to stop them freezing up. With the small figures of the workers beneath the huge black shape in the mist, which was red from the many fires, it gave the impression that some unearthly ritual was being enacted on this bleak southerly bend in the Thames.

Brunel was coming to the conclusion that the pulling power he had hoped to obtain from the barges and tugs on the river was not going to be enough and that his best hope lay in providing much more power to push the ship into the river. The railway engineer Robert Stephenson, a friend who had come to view the operation agreed, and orders were placed with a Birmingham firm, Tangyes Bros., for hydraulic presses capable of much more power.

The whole country was following Brunel's efforts to launch the reluctant ship, with the press increasingly critical. "Why do great companies believe in Mr. Brunel?" scoffed the *Field*. "If great engineering consists in effecting huge monuments of folly at enormous cost to share holders, then is Mr. Brunel surely the greatest of engineers!" There was no shortage of letters offering diverse advice. One critic thought the best plan was to dig a trench up the bows and then push the ship in. Another claimed that 500 troops marching at the double round the deck would set up vibrations that would move the vessel. Yet another idea was to float the ship to the river on cannon balls, or even shoot cannon balls into the cradles. Scott Russell, too, aired his theories on just why the ship was "seizing up" and reluctant to move. He suggested the two moving surfaces should have been wood, not iron.

December and January were bitterly cold. By day the ship looked mysterious in dense fogs; the nights were black as the river itself. Brunel stood alone against a background of criticism; his sheer unremitting determination to get the ship launched permeated every impulse. By early January, he had acquired 18 presses, and they were placed nine at each of the cradles. It was thought that their combined power was more than 4,500 tons.

The new presses were so successful that as the month advanced, the Thames water was lapping her hull. The next high tide of the thirtieth was set for launch day. But the night of the twenty-ninth brought sheeting horizontal rain and a strong southwesterly wind. Brunel knew that if they got the ship launched, the difficulties of managing the craft in the shallow waters of the Thames, where at this point it was not much wider than the length of the ship, would be considerable in such high winds. Miraculously, January 31 was still and calm. The Thames shimmered like a polished surface.

At first light, Brunel started the launch process in earnest. Water that had been pumped into the ship the day before to hold her against the strong tide was now pumped out. The bolts were removed from wedges that were holding the ship. Nothing more could be done until the tide came up the river, which it did with surprising speed and force. Messengers were sent with desperate urgency to collect the men in order not to miss the opportunity, which seemed to have arrived at last. The presses were noisy with effort, and the great Tangyes rams hissed and pushed at the massive structure. After two hours of shoving and straining down the last water-covered part of the ways, the vast iron stern was afloat. The forward steam winch hauled and, quickly, the huge bow responded and moved with solemn deliberation into the water. There were no crowds to witness this defining moment; just a few curious onlookers there by accident watched as the colossal ship moved from one element to another. As the news spread, bells rang out across London as the *Great Eastern* floated for the first time.

Brunel, who had not slept for 60 hours, boarded with his wife and son, and at last could feel the movement of the Great Ship as she responded to the currents of the Thames beneath her. Four tugs took the *Great Eastern* across the river to her Deptford mooring where she could now be fitted out. The cost of the launch was frightening; some estimates suggested as much as £1,000 per foot—and the ship had so far consumed £732,000, with Brunel putting in a great deal of his own money. But the cost to Brunel's health was higher still. Over the past few months he had pushed himself to the limits of endurance. It seemed the Great Ship owned him in body and soul and gave him no relief from the endless difficulties of turning his original vision into a reality. Now that the *Great*

Eastern was finally in the water after *years*, Brunel's doctors insisted that he take a rest.

When Brunel returned in September 1858, he found that the Eastern Steam Company was in debt, that there was no money to fit out the ship, and there was talk of selling her. The company tried to raise £172,000 to finish the work on the ship, but this proved impossible. The financial problems of the board were only resolved when they formed a new company, The Great Ship Company, which bought the *Great Eastern* for a mere £160,000, and allotted shares to shareholders of Eastern Steam in proportion to their original holdings. Brunel was reengaged as engineer and, with his usual energy, became busy with designs for every last detail, even the skylights and rigging. But now he was harassed with health problems; doctors had diagnosed his recurring symptoms as Bright's disease, with progressive damage to his kidneys. They insisted that he spend the winter relaxing in a warmer climate. The last thing Brunel wanted was to leave his Great Ship when there was still so much to supervise. Reluctantly he agreed to travel to Egypt with his wife and son, Henry.

Before he left, with memories of the impossible position that the board had faced when dealing with Scott Russell, Brunel urged them to ensure that any contract they entered into for fitting out the ship be absolutely binding. But with Brunel abroad and clearly unwell, the board, left with bringing to completion such a unique vessel, opted for the devil they knew. Scott Russell had built the hull and he was building the paddle engines. He was, after Brunel, the man who knew most about the Great Ship. Thus, it wasn't long before the charming, charismatic Scott Russell with his delightfully low-priced, somewhat ambiguous contract was back onboard.

In May 1859, Brunel returned. The enforced holiday appeared to have been beneficial, and his friends were hopeful that he was fully recovered. Privately he knew this was not the case. His doctors had made it quite clear: his disease was progressing relentlessly. Only so much time was left for him and he should certainly not overexert himself. Yet for Brunel, rest was out of the question. Whatever private bargain he may have made with himself, it proved impossible for him to resist the pull of the *Great Eastern*. His ship came first, whatever the cost.

With the maiden voyage planned for September, all Brunel could see was the enormous amount still needing to be done. So he rented a house near the ship and, with his usual energy, dealt daily with the many problems that needed his expert attention. Everything from the engine room to the rigging was checked; the best price of coal ascertained; the crew for the sea trials named; progress reports on the screw engine prepared; notes for Captain Harrison written; advice on the decoration in the grand saloon given—nothing escaped his practiced eye. On August 8, 1859, a grand dinner was given for MPs and members of the House of Lords in the richly gilded rococo saloon, but Brunel was too exhausted to attend. It was a glamorous occasion, and in Brunel's absence, Scott Russell rose to it, shining in the glowing approbation of the distinguished audience.

On September 5, Brunel was back onboard his ship. He had chosen his cabin for the maiden voyage and stood for a moment on deck by the gigantic main mast while the photographer recorded the event. He had lost weight; his face was thinner; and his clothes hung on him. In one hand he held a stick to help him get about; his shoes were clean and polished. He had a fragile, expended air. As he looked at the camera—his eyes, as always concentrating, absorbed in some distant prospect—his image was of a man with little time left. Just two hours later he collapsed with a stroke. He was still conscious as his colleagues carried him as though he were breakable to his private coach and slowly drove him home to Duke Street.

THE *GREAT EASTERN* made her way alone now, without Brunel's attention, directed by fussing tugs, down the Thames to Purfleet in Essex and beyond for her sea trials. Once out to sea, she was magnificent. "She met the waves rolling high from the Bay of Biscay," reported the *Times*. "The foaming surge seemed but sportive elements of joy over which the new mistress of the ocean held her undisputed sway."

With Brunel lying paralyzed at home on Duke Street, Captain Harrison was now in charge, but his command was diluted as the engine trials took place. The two engine rooms were supervised by the representatives of the firms responsible for the engines; Scott Russell put his man Dixon in charge of the paddle engine room. Also problematical was that Brunel had designed many new and innovative features for the smoother running of his ship, some of which those in charge were neither familiar with nor were even aware existed. No one had the intimate understanding of the complex workings of the ship that Brunel had and if his familiar figure had been onboard, in total charge, calling on his boundless energy and sharp mind to direct proceedings, it is possible the accident would never have happened.

The *Great Eastern* was steaming along, just off Hastings, cutting smoothly through the waves and making light work of the choppy seas that were tossing the smaller boats dangerously around her. Passengers onboard had left the glittering chandeliered saloon to dine; a hardier group had gathered on the bow to view the distant land. Suddenly, without warning, there was a deafening roar that seemed to come from deep within the bowels of the ship, and the forward funnel was wrenched up and shot 50 feet in the air, accompanied by a huge cloud of steam under pressure. Up in the air was tossed the forward part of the deck and all the glitter and glory of the saloon in the catastrophic eruption. Passengers were stunned, almost blinded by the white cloud of steam, and then the debris came crashing down.

Captain Harrison seized a rope and lowered himself down through the steam into the wreck of the grand saloon. He found his own little daughter, who by a miracle had escaped unhurt. The eruption, it was clear, had occurred in the paddle boiler room, which had suddenly been filled with pressurized steam. Several dazed stokers emerged on deck with faltering steps, their faces fixed in intense astonishment, their skins a livid white. "No one who had ever seen blown up men before could fail to know that some had only two or three hours to live," reported the *Times*. "A man blown up by gunpowder is a mere figure of raw flesh, which seldom moves after the explosion. Not so men who are blown up by steam, who for a few minutes are able to walk about, apparently almost unhurt, though in fact, mortally injured beyond all hope of recovery." Since they could walk, at first it was hoped that their injuries were not life-threatening. But they had met the full force of the pressurized steam; they had effectively been boiled alive. One man was quite oblivious to the fact that deep holes had been burned into the flesh of his thighs. A member of the crew went to assist another of the injured and, catching him by the arm, watched the skin peel off like an old glove. Yet another stoker running away from the hell below, leapt into the sea, only to meet his death in the blades of the paddle wheel.

"A number of beds were pulled to pieces for the sake of the soft white wool they contained," reported *Household Words*, "and when the half-boiled bodies of the poor creatures were anointed with oil, they were covered over with this wool and made to lie down. They were nearly all stokers and firemen, whose faces were black with their work, and one man who was brought in had patches of red raw flesh on his dark, agonised face, like dabs of red paint, and the skin of his arms was hanging from his hands like a pair of tattered mittens. . . . As they lay there with their begrimed faces above the coverlets, and their chests covered with the strange woolly coat that had been put upon their wounds, they looked like wild beings of another country whose proper fate it was to labour and suffer differently from us." Amongst the crew, the fate of the riveter and his boy, rumored to have been locked alive in the hull years before, resurfaced, with prophecies of worse to come.

Still, the Great Ship sailed confidently on, her engines pounding, not even momentarily stopped in her tracks, just as her designer envisaged. Such an accident would have foundered any other ship. The stunned passengers were at a loss to know what had happened, and were totally unaware that the whole explosive performance could be repeated at any minute in the second funnel. The accident had been caused by a forgotten detail: a stopcock had been inadvertently turned off, causing a water jacket in the forward funnel to explode under pressure. There was a similar stopcock affecting a second funnel, also switched off, also building up enormous pressure. By a miracle, one of Scott Russell's men from the paddle engine room realized what had happened and sent a greaser to open the second stopcock. A great column of steam was released and the danger passed.

Twelve men had been injured; five men were dead. An inquest was held in Weymouth in Dorset. It proved difficult to ascertain just who had responsibility for the stopcock. Scott Russell, smoothly evasive, claimed that he had been onboard merely in an advisory capacity and that the paddle engines and all connected with them were the total responsibility of the Great Ship Company. "I had nothing whatever directly or indirectly to do," he said. "I went out of personal interest and invited Dixon as my friend." He claimed he had "volunteered his assistance only when it became obvious to him that the officers in charge were having difficulty in handling the ship." All the other witnesses also disclaimed responsibility. There were, however, passengers on the ship who insisted that they had heard Scott Russell, "give at least a hundred orders from the bridge to the engine room." The truth, though, was undiscoverable, lost somewhere in the maze of evidence from the variously interested parties. Accidental death was the verdict from the jury.

At his home in Duke Street, Brunel, a shadow, but still clinging to life, was waiting patiently through the dull days for word of the sea trials. Paralyzed, he lay silently, hoping for good news, hoping so much to hear of her resounding success. Instead, he was told of the huge explosion at sea and the terrible damage to the ship. It was a shock from which he could not recover. It was not the right news for a man with such a fragile hold on life. Nor was it just payment for his labor of love. He died on September 15, just six days after the explosion on his beloved *Great Eastern*, still a relatively young man of 53.

The country mourned his loss; the papers eulogized. A familiar brilliant star was suddenly out. "In the midst of difficulties of no ordinary kind," said the president of the Institution of Civil Engineers, Joseph Locke, "and with an ardour rarely equalled, and an application both of body and mind almost beyond the limit of physical endurance in the full pursuit of a great and cherished idea, Brunel was suddenly struck down, before he had accomplished the task which his daring genius had set before him."

With her creator gone, the Great Ship put in to Weymouth on the south coast for repairs that would bring her up to Board of Trade requirements. While work was in progress, sightseers at two shillings and sixpence per head came in the thousands to wonder and whisper if the ship was damned. Was there not a story about a riveter and his boy and bad luck? Since the repairs took longer than expected, the company postponed plans for the maiden voyage until the following spring; and because the sightseers had proved financially successful, the Great Ship steamed on to Holyhead in Anglesey, Wales, to find more.

While there she encountered a fearsome storm of gale-force winds and rivers of rain. The crashing waves broke the skylights, deluged the ship, and ruined the décor in the grand saloon. Captain Harrison realized she had lost her mooring chains and was at the mercy of the storm. He ordered the engineer to start the paddle engines and, by skillfully keeping her head into the wind, the *Great Eastern* rode out the storm that saw many wrecked around her. Nearby the steamship *Royal Charter* went down, claiming 446 lives, further evidence that the *Great Eastern* was unsinkable, as her creator had claimed. Nevertheless, it was decided that Southampton would be a more sheltered port for the winter.

It was, however, a troubled winter for the *Great Eastern* and those connected with her. At a special shareholders meeting, the news that there was a mortgage of £40,000 attached to the ship, and that the company was over £36,000 in debt, led to angry calls for the directors to resign. A new board of directors was voted in, led by Brunel's friend, Daniel Gooch. They issued shares in the hopes of raising £100,000 to complete the ship, but doubts were raised over the terms of the contract with Scott Russell and, above all, how £353,957 had been expended on a ship still not fit for sea. Once again, Scott Russell's loosely defined estimate for fitting out the ship had been wildly exceeded. His services were finally dispensed with, and Daniel Gooch was elected chief engineer.

Yet more bad luck seemed to haunt those connected with the ship. One January day in the winter of 1860, Captain Harrison and some members of the crew set out from Hythe pier by the Solent to reach the *Great Eastern* in a small boat. As they left the shelter of the land, they were hit by a sudden violent squall. The tide was very high, with choppy, dangerous seas. Harrison ordered the sail down, but the wet sail would not budge. In an instant, the wind hit the sail and turned the boat over. Captain Harrison, always cool and collected, made repeated attempts to right the boat, but it kept coming up keel first. Buffeted by wild and stifling seas, he seemed almost powerless. Very quickly, the sea claimed the ship's boy, the coxswain, and the master mariner, Captain Harrison himself. The *Great Eastern* had lost her captain, handpicked by Brunel.

Almost two years had elapsed since the launch and the ship still had not made a trip to Australia as originally planned. The continued shortage of funds made this difficult to finance, so instead the Great Ship Company decided to sail her as a luxury liner on the Atlantic. Finally, in June 1860, the *Great Eastern* sailed for New York on her maiden voyage. Onboard were just 38 passengers, who marveled at the great ship, fascinated by the massive engines, the imposing public rooms, the acres of sail. Despite the small number of people onboard, they were in a champagne mood, enjoying dancing, musicals and a band, and strolling the wide deck as they seemed to walk to America with hardly a roll from the *Great Eastern*. The crew, 418 strong, took her across the Atlantic as though she were crossing a millpond.

"It is a beautiful sight to look down from the prow of this great ship at midnight's dreary hour, and watch the wondrous facility with which she cleaves her irresistible way through the waste of waters," reported the *Times*. "A fountain, playing about 10 feet high before her stem, is all the broken water to be seen around her; for owing to the great beauty of her lines, she cuts the waves with the ease and quietness of a knife; her motion being just sufficient to let you know that you have no dead weight beneath your feet, but a ship that skims the waters like a thing of life. . . ."

When she reached New York, on June 27, people turned out in the thousands to view her; they packed the wharfs, the docks, the houses. "They were in every spot where a human being could stand," observed Gooch. A 21-gun salute was fired and the ecstatic crowds roared their approval. All night in the moonlight, people tried to board her, and the next day an impromptu fairground had formed, where *Great Eastern* lemonade and oysters and sweets were sold. Over the next month, the streets were choked with sightseers, and the hotels were full as people were drawn from all over America to view this great wonder. The *Great Eastern* was the toast of New York. Her future as a passenger ship seemed assured.

IN THE MID-NINETEENTH century, sea travel was hazardous; many ships were lost, but the *Great Eastern* was proving to be unsinkable. Confidence in her was rising. She had survived the terrible "*Royal Charter* storm," and with her double hull—thought to be longer than the trough of the greatest storm wave—and transverse and longitudinal watertight bulk-heads, she would "laugh" at rough weather.

And so it was with a mood of great optimism that the *Great Eastern* left Liverpool under full sail on September 10, 1861, in such soft summer weather that some of her 400 passengers were singing and dancing on deck. The next morning was gray, with a stiff breeze blowing spray. By lunchtime, they were meeting strong winds and heavy seas, and by the afternoon, when they were 300 miles out in the Atlantic, the winds were gale-force. "She begins to roll very heavily and ship many seas," wrote one anxious passenger. "None but experienced persons can walk about. The waves are as high as Primrose Hill."

With waves breaking over the ship, the *Great Eastern* soon leaned to such an extent that the port paddle wheel became submerged. The captain, James Walker, became aware of a sound of machinery crashing and scraping from the paddle wheel. Clinging onto the rails, he lowered himself towards the ominous sounds. As the Great Ship lurched through the waves and the paddle wheel was flung underwater, Walker was pulled into the sea up to his neck. When he got a chance to investigate, he could see planks splitting, girders bending, and the wheel scraping the side of the ship so badly that it looked as though it might hole her. Somehow he clambered back to the bridge and gave the order to stop the paddle engines. But without these, the screw engine alone could not provide enough power to control the ship's movement in what was fast turning into a hurricane.

As the ship rolled, the lifeboats were untied and thrown into the foaming water by the furious waves. One lifeboat near the starboard paddle wheel was hanging from its davit and dancing about in a crazy fashion. It was damaging the paddle wheel, so Captain Walker had it cut away, then ordered the starboard engines reversed to ensure the rejected boat did not harm the paddle further. The ship was wallowing and plunging at an angle of 45 degrees in the deep valleys of water, with over-

whelming mountainous seas on either side. Walker was desperate to turn
the ship into the wind, but before he could, the paddle wheels were swept
away by a huge wave.

Terrifying sounds were coming from the rudder, which had been
twisted back by the heavy seas. It was out of control and was crashing
rhythmically into the 36-ton propeller, which was somehow still turning.
With the loss of steerage from the broken rudder it was quite impossible
to turn the ship. And then the screw engine stopped. The ship was now
without power and completely at the mercy of the elements. An attempt
was made to hoist some sail, but the furious gale tore them to shreds.

Belowdecks, madness reigned. In the grand saloon, chairs, tables, and
the grand piano were being smashed to pieces. Rich furnishings were
shredded. The large cast-iron stove came loose and charged lethally into
mirrors, showering fragments of glass everywhere and snapping the elegant decorative columns in half. Amongst this chaos were passengers trying to dodge the furniture and cling for dear life to something stable.
Water roared and poured in torrents through the broken skylights. At
one moment, two cows from the cowpen on deck and some hens were
thrown in with the deluge. A swan, which was trying to take flight, battered itself to death. The cabins were smashed and soaked as water
poured belowdecks. The ship's doctor was busy treating 27 passengers
with serious fractures, including the ship's baker, whose leg had been
broken in three places.

For the whole of the following day the hurricane continued to blow
and the *Great Eastern* took on even more water. A four-ton spar loaded
down with iron was put overboard in the hopes it would act as a drag and
give some control to the seemingly doomed ship. It was soon torn away.
By now, water was coming in through portholes and skylights and finding its way belowdecks, where it overwhelmed the pumps. Ominous
thumping and crashing was heard from the holds where nothing had been
secured, and ruined goods were flung from side to side, with the water in

the hold battering the hull. No one had eaten for 48 hours. The situation looked hopeless.

In the evening an attempt was made to secure the wildly gyrating rudder, still battering itself uncontrollably into the propeller and gradually being torn apart. Heavy chains were eventually wound around the broken steering shaft and made fast, so that the rudder at last became stationary. A sailor volunteered to brave the seas, and was lowered on a boatswain's chair to the rudder where, in spite of terrifying conditions, he managed to loop a chain around the rudder and through the screw opening. At last, with two chains attached to the rudder, a very primitive means of steering was established. It was now possible to start the screw engines and slowly make it back to Milford Haven in Wales.

When land finally came into view, the passengers, wild with excitement, began an impromptu celebration in the ruined saloon. Limping into harbor with the band playing, the *Great Eastern* had survived. Everyone onboard was quite sure no other ship could possibly have done so. She was indeed unsinkable. The grateful passengers disembarked, one woman so overwhelmed that she fainted. But the bill for the damage, which took months to repair, was £60,000. The company was in debt again.

By the summer of 1862, the *Great Eastern* was making regular, successful, and uneventful trips to New York, with a new captain, Walter Paton. In August, she embarked once again with a full cargo and 1,500 passengers; and although the ship met with yet another bad storm, Captain Paton stayed on the bridge and battled through it at full power. They reached New York on a night of calm waters and silver moonlight where the ship proceeded through the narrow channel to dock. During this final length of the journey off Long Island, a deep rumbling sound was heard; the ship faltered and then recovered. No damage could be found. Next day at anchor, the captain sent a diver down, who discovered a great gash, 85 feet long and 5 feet wide, on the flat bottom of the outer hull.

The inner hull was untouched. Brunel's double hull had saved the ship. She had suffered this awful wound from an uncharted needle of rock that came within 25 feet of the surface.

Captain Paton was a long way from Milford Haven, where the ship could be put on a grid iron for repairs. He did not want to chance a journey back across the Atlantic at a time of equinoxial gales with a hull so badly ruptured, but there was no facility in America where the ship could be repaired, and no dry dock big enough to take her. Even if he could beach her massive hull somewhere in North America, the long gash was in the flat bottom of the ship, hence would be impossible to repair.

Though ship builders in North America were intrigued by the challenge of repairing the vast ship, no one had an answer until a civil engineer named Edward Renwick, and his brother, Henry, offered their services. Unfortunately, neither man inspired confidence. Both had only partial sight and were inclined to grope their way around furniture. They, however, were confident that, in spite of their disabilities, they could repair the ship. Their plan was to build a watertight cofferdam over the long gash, enabling work to be completed in the dry.

Their plan was accepted, and templates were made from the inner hull. The space between the hulls was calculated, which then gave the exact shape of the outer hull. The riveters were to conduct their job from inside the ship by making their way down a dark shaft to the gash in the hull—though many needed some persuading to trust the temporary cofferdam clamped on to the big ship's hull. One day panic ensued when knocking was heard in the double hull. Rumors spread quickly and the riveters became adamant that a ghost was hammering. They dropped their tools and refused to work as long as the banging continued.

The captain was called. He, too, heard the supposed ghost "pounding on the hull." Work was stopped. Fear infiltrated the ship like mist. Every inch of the bilge was inspected. The hammering was coming from below the waterline so Captain Paton inspected the outside of the hull in a small boat. There the "ghost" was discovered: a loose chain knocking the side of the ship as it rose and dipped in the swell.

The work was finished in December 1862, and Edward and Henry Renwick presented their bill for £70,000. The insurance firm refused to pay. The company had now lost £130,000 in the last two years. When the *Great Eastern* arrived back in England, she spent months on a grid iron while Board of Trade inspectors reviewed the work of the Renwick brothers. Subsequently, she made three more trips across the Atlantic, lost another £20,000, and was then beached again while the board considered the situation. Despite their efforts, the board had failed to make their fortunes from the unique vessel; even while on her grid iron in some lonely cove, like a great sea creature thrown up onto the beach and forgotten, she was still silently absorbing funds. The company decided to sell their one asset. The *Great Eastern* was auctioned on January 1864 for the disappointing sum of £25,000.

Far from being finished, however, the *Great Eastern* was on the threshold of a completely new career. The chairman of the new company was Daniel Gooch, who had always believed in Brunel's Great Ship. He immediately chartered the *Great Eastern* to the Atlantic Telegraph Company for £50,000 of cable shares, which intended to use it to lay cable across the ocean from Ireland to America. There had already been an unsuccessful attempt to accomplish this feat by a wealthy American businessman, Cyrus Field, who was quite sure that it was possible to make the cable link between the two continents because the sea floor between Newfoundland and Ireland was plateaulike and not too deep.

Thus, the *Great Eastern* was stripped of all her finery and prepared for the cable laying. The grand saloons and palatial first-class cabins were renovated to house the miles of cable and the machinery that would deliver it to the ocean. And in July 1865, with Gooch onboard, she began her next venture.

Gooch had never lost faith in the ship in spite of the bad luck and bad management that had been her lot. He knew Brunel had designed a strong and magnificent ship, one that had come through adversity time and again. He felt her worth would at last be realized. The success of the venture, he declared, "will open out a useful future for our noble ship, lift her out of the depression under which she has laboured from her birth and satisfy me that I have done wisely in never losing confidence in her."

Despite his enthusiasm, he too soon ran into problems. On August 2, 1865, after successfully laying 1,000 miles of cable across the Atlantic, the cable broke and disappeared 2,000 fathoms into the faceless ocean, which offered no clues or help. "All our labour and anxiety is lost," despaired Gooch. "We are now dragging to see if we can by chance recover it, but of this I have no hope, nor have I heart to wish. I shall be glad if I can sleep and for a few hours forget I live. . . . This one thing upon which I had set my heart more than any other work I was ever engaged on, is dead." The ship returned, defeated, having lost £700,000 worth of cable.

From defeat, optimism once again blossomed, and in July 1866, after reviewing the mistakes of the previous year, the ship sailed again with stronger cable and improved machinery. This time success was the reward as the ship put in to Hearts Content Bay, Newfoundland. Gooch sent a telegram back to the Old World: "Our shore end has just been laid and a most perfect cable. . . ." And when, in September, returning home they reached the approximate position of the previous year's lost cable, they put the improved grappling gear to work, and by some small miracle found and recovered it. Success now seemed assured, and Gooch estimated that the company would be nearly £400,000 a year better off. For the next three years, the *Great Eastern* laid cable all over the world: from France to America, Bombay to the Red Sea, and a fourth cable to America, as well as completing repair work on previously laid cable. In 1874, however, her cable-laying days were brought to an end with the launch of a custom-built ship, the *Faraday*, which had been produced especially for the task of cable-laying.

Now the *Great Eastern* presented the company with a problem. No one had ever made money from her as a passenger ship. She had been designed originally to steam halfway round the world to Australia and back, where, with 4,000 passengers and enough fuel for the return journey, she would have been strong competition for sailing ships and made a fortune. But the Suez Canal was now in operation and the *Great Eastern* was too large to use it. Any journey she now made to Australia would not be competitive, and she was always considered too large to be economic on the Atlantic run.

The ship spent 12 quiet years, largely forgotten, on the grid iron at Milford Haven, while a gentle dilapidation settled on her like a mold. Suggestions were made for her future, perhaps as a hospital or hotel, but these came to nothing. She was auctioned in 1885 for £26,000 to a coal hauler, Edward de Mattos. He leased her for a year to a well-known Liverpool draper, Louis Cohen, who proposed to turn her into a showboat for the exhibition of manufacturers.

The old ship was goaded into life, but by that time her paddle engines were eaten up with rust and unusable. Even her screw engines were stiff and reluctant as she slowly made her way to Liverpool, to be turned into some sort of funfair decked with advertisements. Trapeze artists dived from the rigging. The saloon became a music hall. Cheap fairground attractions vied with beer halls, conjurers, knife throwers, all the fun of the fair brought in a profit for Cohen. "Poor old ship, you deserved a better fate," Gooch wrote in his diary on hearing the news, adding, "I would much rather the ship was broken up than turned to such base uses."

When Cohen's lease expired, de Mattos tried to repeat his success, but it seems he did not have enough of the circus in his blood to pull it off. So, in 1888, after due consideration, the ship that had cost well over £1 million pounds to build, maintain, and repair, was auctioned for scrap for the meager sum of £16,000.

The still-magnificent ship and all the dreams she carried for everyone connected with her was taken to the scrapyard at Birkenhead to be demolished. At first it looked as though the brakers would make a tidy profit. They estimated that they could sell the iron plates and various metals for £58,000; but with the *Great Eastern*, making a profit was never a foregone conclusion; and true to her history, she made a loss. Demolition proved cripplingly expensive, as human hands were not enough to pick the immensely strong hull to pieces. This was a ship designed for strength by her creator, a ship that had survived the full fury of Atlantic storms. It took some 200 men working night and day for two years, swinging demolition bells and anything else they could find to poleax her into so many tons of scrap. Slowly, the layers of metal were peeled away—the outer skin of the hull, the inner hull, the organs of the engine, and all the intimately connecting shafts and pistons—until one day, where she had stood was just space.

As for the riveter and his boy said to have been entombed alive in the double hull, rumors spread that their skeletons were indeed found. According to James Dugan, author of *The Great Iron Ship* written in 1953, there was one witness: a Captain David Duff, who at the time was a cabin boy. He claimed to have visited the wrecker's yard and wrote: "They found a skeleton inside the ship's shell and the tank tops. It was the skeleton of the basher who was missing. Also the frame of the bash boy was found with him. And so there you are Sir, that's all I can tell you about the *Great Eastern*." But the local papers of the time bear no record of this extraordinary story, and the captain's account has never been authenticated. To date, the mystery of the entombed basher and his mate remains unsolved.

It is wrong to blame the ill fortune that seemed to haunt the *Great Eastern* on some evil-spirited ghosts. Her design was way ahead of its time: she was to remain the largest ship in the world until the *Lusitania* of 1906 and, later, the *Titanic*. But in the mid-nineteenth century, there were few harbors where the *Great Eastern* could dock, and this fact alone limited her success. She was specifically designed for taking large numbers of people to Australia, but this never happened. Those who managed her have been criticized for their insistence on using her for the luxury market to the States at a time when there was just too much competition on this route. If she had sailed to New York from Liverpool with a full complement of emigrants and brought back cotton or wheat, she could have made £45,000 per round-trip. More than 800,000 emigrants left Europe for the States in the Civil War years.

Had her creator lived beyond the age of 53, perhaps he would have been able to steer his ship towards profitability. But Isambard Kingdom Brunel was gone, and in 1890, his dream disappeared, too. Only the spirit of the man lived on. His friend Daniel Gooch wrote on Brunel's death, "The greatest of England's engineers was lost, the man with the greatest originality of thought and power of execution, bold in his plans but right. The commercial world thought him extravagant; but although he was so, great things are not done by those who sit down and count the cost of every thought and act."

—

CHAPTER 2

THE BELL ROCK
LIGHTHOUSE

There is not a more dangerous situation upon
the whole coasts of the Kingdom, or none that calls
more loudly to be done than the Bell Rock . . .

ROBERT STEVENSON, *1800*

THE SAFE ANCHORAGE of the Firth of Forth on the east coast of
Scotland has always been a refuge for ships hoping to escape the wild
storms of the North Sea. The safety of this natural inlet, however, is considerably compromised by the presence of a massive underwater reef, the
Bell Rock, placed treacherously in the middle of the approach to the
Firth of Forth. It is far enough away from the coast so that landmarks are
unable to define its position, being 11 miles south of Arbroath and a similar distance west from the mouth of the Tay. At the beginning of the
nineteenth century, when a storm was brewing in the Forth and Tay area,
those at sea faced a forbidding choice: ride out the storm in the open sea
or try to find safety in the Firth of Forth and risk an encounter with the
Bell Rock.

Hidden by a few feet of water under the sea, the craggy shape of the
Bell Rock lay in wait for sailing ships, as it had for centuries, claiming
many lives and ships and scattering them wantonly, like trophies over its

Patterns for the stones of one course

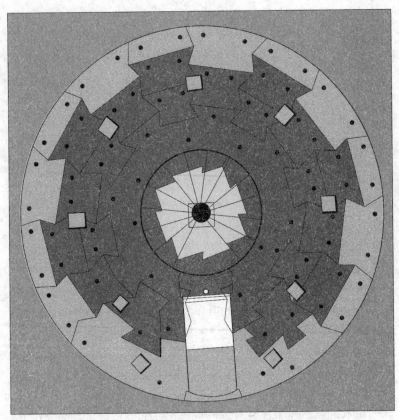

One of the courses of stones, showing how the blocks
were dovetailed into one another

silent and mysterious escarpments. It bares itself briefly twice a day at low tide for an hour or two, and then disappears under the sea at high tide, its position sometimes given away by waves breaking on the submerged rocks and foaming surf over its rugged features. An outcrop of sandstone about a quarter of a mile long, it slopes away gently on the southern side, but to the north it rises sheer from the seabed, an unyielding barrier.

For early navigators the greatest danger was to come suddenly upon the northern cliff face. Any ship taking soundings north of the rock would find deep water and assume all was safe, only to learn the fatal error should the ship stray a few yards farther south. All onboard would listen for the last sounds they might hear of timber being torn and split as wood was crushed against rock. So many lives were lost along the whole Scottish coast that, according to a local report, the notorious Bell Rock "breathed abroad an atmosphere of terror."

For centuries, the sea-lanes were deserted, their wild highways left unchallenged, but from about the mid-eighteenth century the growth of trade in flax, hemp, and goods for the weaving industry saw an increase in shipping and, as a consequence, a growing number of fatal collisions with the massive submerged cliff of the Bell Rock. The heavy toll brought pleas for some kind of warning light, although no one was sure how this could be done so far out to sea on a rock that for most of the time was underwater.

The local people of the east coast had once succeeded in putting a warning on the rock. In the fourteenth century it was said a man named John Gedy, the abbot of Aberbrothock, was so concerned at the numbers who perished there that he set out to the rock with his monks and an enormous bell. With incredible ingenuity, they attached the bell to the rock and it rang out loud and clear above the waves warning all seafarers, like an invisible church in the sea.

The good abbot, however, had not reckoned on human avarice. Soon after, a Dutch pirate called Ralph the Rover stole the bell, in spite of its power to save life by its insistent warning ring. Ironically, he died within a year and must have regretted his act when his ship met bad weather, the great reef, and, some said, a deserving fate, as he and his ship disappeared beneath the waves. From that time, the rock became known as the "Bell Rock."

The coast of Scotland is long and rugged and has many jagged peninsulas and rocky islets. Even by the late eighteenth century, for hundreds of miles, according to local accounts, these desolate shores "were nightly plunged into darkness." To help improve the safety of these coastal waters, the Northern Lighthouse Board was established in 1786 to erect and maintain lighthouses. At that time, the warning lights to ships were often no more than bonfires set on dangerous headlands, maintained by private landowners. But in bad weather, when the warning fires were most needed, they were usually put out by drenching rain.

By 1795, the board had improved on these primitive lights with the construction of seven major lighthouses, but progress was slow. They were chronically underfunded, though never short of requests to do more by worried shipowners, especially to put a light on the Bell Rock. The Northern Lighthouse Board was well aware of the need for a light on the rock. Its reputation as a killer lying in wait at the entrance to the enticing safety of the Firth of Forth had traveled well beyond England. However, with little in the way of funds and the difficulties of building so far out to sea on a rock that was submerged in up to 16 feet of water for much of the day, such a request was seen as improbable madness, not even to be considered.

There was one man, however, who had been dreaming of the impossible, of building a lighthouse on the hidden reef, thereby enabling the whole bay of the Firth of Forth to be useful as safe anchorage. Robert Stevenson was a man of strong character, who by some strange fate had been given the very opportunities he needed to fulfill his ambition—

though in early life, his chances of success had looked poor. His mother, Jean Stevenson, had been widowed and left penniless when he was only 2. Years of hardship followed, but Jean Stevenson, a deeply religious woman, struggled to ensure an education for her son. In later life, Stevenson always remembered "that dark period when my mother's ingenious and gentle spirit amidst all her difficulties never failed her."

Jean eventually remarried, in November 1792, to an Edinburgh widower named Thomas Smith, who designed and manufactured lamps. At the time, Smith was interested in increasing the brightness of his lamps. A scientific philosopher from Geneva, Ami Argand, had recently developed a way of improving brightness by fitting a glass tube, or chimney, around the wick. Smith was experimenting with taking this work further by placing a polished tin reflector behind and partly surrounding the wick, shaped in a parabolic curve to focus the light. This gave a much brighter beam than conventional oil lamps, and the lamps from his workshops were now much in demand. He was soon approached by the Northern Lighthouse Board, which employed him as their lighting engineer.

When the young Robert Stevenson visited his stepfather's workshop, he found it a magical place where uninteresting bits of metal and glass were transformed into beautiful precision-made objects. Jean could see where her son's interests lay and, much to his delight, Stevenson was soon apprenticed to Smith. One of Smith's duties at the Northern Lighthouse Board was to visit the board's growing number of lighthouses. During the summer months he and Stevenson would set out by boat and appraise the situation, repairing damage and deciding on the position of new lighthouses. By around the turn of the century, this responsibility fell entirely to Stevenson.

"The seas into which his labours carried the new engineer were still scarce charted," his grandson, Robert Louis Stevenson, wrote years later. "The coasts still dark; his way on shore was often far beyond the convenience of any road; the isles in which he must sojourn were still partly savage. He must toss much in boats; he must often adventure on horseback through unfrequented wildernesses; he must sometimes plant his lighthouse in the very camp of wreckers; and he was continually enforced to the vicissitudes of out door life. The joy of my grandfather in this career was strong as the love of woman. It lasted him through youth and manhood, it burned strong in age and at the approach of death his last yearning was to renew these loved experiences."

From May to October, Stevenson went on his rounds visiting the board's scattered lighthouses, taking much-needed supplies and solving problems. The latter could vary from the repair of storm-damaged buildings to finding new pasture for the keeper's cow. Stevenson, also employed to map out the position of new lighthouses, soon found that some of the inhabitants of the remote islands—who supplemented their income from wrecking—were openly hostile to him.

On one journey in dense fog, his ship came dangerously near sharp rocks of the Isle of Swona. The captain hoped to get help towing the ship away from the danger from a village he could see on shore. The village, however, looked deserted; everyone was asleep. To attract attention, he fired a distress signal. Stevenson then watched in disbelief, as "door after door was opened, and in the grey light of morning, fisher after fisher was seen to come forth nightcap on head. There was no emotion, no animation, it scarce seemed any interest; not a hand was raised, but all callously waited the harvest of the sea, and their children stood by their side and waited also." Luckily a breeze sprang up and the ship was able to make for the open sea.

During these summer trips Stevenson learned a great deal. He could be impatient, and was not inclined to suffer fools gladly, but he never lacked confidence in his ability to tackle the most difficult problems. Over these years, as the Scottish coastline and its lighthouses became ingrained on his mind, he was nurturing his secret ambition to tame forever the awful power of the Bell Rock. But the fulfillment of his dream seemed remote. Stevenson was not a qualified civil engineer; and as Smith's young assistant he had little influence with the board. And he was only too aware that the commissioners believed that a light on the Bell Rock was out of the question.

Those living on the northeast coast of England and Scotland in December 1799 saw the old century dragged out in a thunderous storm of screaming winds and mountainous seas, which raged from Yorkshire to the Shetlands. All along the east coast, ships at anchorage were torn from their moorings and swept away. Those seafarers who could hear anything above the wind and crash of waves, would listen for the dreaded sound of wood cracking and splitting as it was thrown against rock: the sound of death. In Scotland, the haven of the Firth of Forth, guarded by the Bell Rock, was ignored. Ships preferred to make for the open sea and take their chances in the storm rather than try and steer their way past the dreaded reef. The storm lasted three days and was to sink 70 ships.

The call for a light on the Bell Rock grew louder. If there had been a lighthouse, shipowners argued, many more ships would have made for the safety of the Firth of Forth. The Northern Lighthouse Board began, at last, to give serious consideration to what they still saw as an insolvable problem, and Stevenson was quick to present his own plan for a beacon-style lighthouse on cast-iron pillars. Although there was not a more dangerous situation "upon the whole coasts of the Kingdom," he argued, his design would be safe, relatively inexpensive, and even pay for itself as the board collected fees from ships taking advantage of its warning light. The cautiously minded board was impressed with the idea of economy, but less sure of Stevenson's design.

Despite his travels around the coast of Scotland, Stevenson had not yet managed to set foot on the Bell Rock itself and was impatient to do so. In April 1800, he hired a boat, intending to survey the site, but the weather was too stormy to land. In May, as he sailed nearby on a journey north, it lay invisible, even at low tide. He had to wait until the spring tides of October before he could make the attempt again. At the last minute, however, the boat he had been promised was unavailable and no one was prepared to take him out to the rock, not even in calm seas. Time was running out for a landing on the rock before winter, and if he could not find a boat, he would miss the favorable tides. Finally, a fisherman was found who was prepared to take the risk—it turned out that the man often braved the Bell Rock to hunt for valuable wreckage to supplement his income.

Once on the rock, Stevenson and his friend, the architect James Haldane, had just two hours in which to assess the possibilities that the rock might offer before the tide returned and the rock disappeared. It was covered in seaweed and very slippery. The surface was pitted, and seawater gurgled and sucked in the fissures and gulleys that crisscrossed the rock; nevertheless, Stevenson was encouraged by what he saw. The exposed area at low tide was about 250 by 130 feet, revealing enough room for a lighthouse. Better still, the surface of the rock was of very hard sandstone, perfect for building.

There was one problem, though. He had thought that a lighthouse on pillars would offer less resistance to the sea, but when he saw the heavy swell around the rock, overwhelming the channels and inlets, pushing their bullying foamy waters into deep fissures even on a calm day, he knew his plan could not work. Visiting boats bringing supplies or a change of keeper would be shattered against the pillars in heavy seas, and the capability of the pillars to withstand the timeless beating of the waves was questionable as well. "I am sure no one was fonder of his own work than I was, until I saw the Bell Rock," he wrote. "I had no sooner landed than I saw my pillars tumble like the baseless fabric of a dream."

The two hours passed all too quickly. As the returning tide swirled around their feet, the fisherman who had gathered spoils from wreckage on the reef became anxious to leave. For Stevenson, finding the Bell Rock and standing at the center of its watery kingdom, with nothing but the ever-encroaching sea in sight, had been a revelation. It was clear to him that only an immensely strong tower would have a chance of surviving in such an exposed position: a building higher than the highest waves, made of solid sandstone and granite. With these thoughts in mind, he undertook an extensive tour of English lighthouses and harbor lights, in search of a model on which to base his own plans. It was a journey of some 2,500 miles by coach or on horseback, which took many months of 1801. He soon found there was only one such stone sea-tower already in existence, built on a buttress of rock about nine miles from the port of Plymouth, off the south coast in Cornwall.

The Eddystone lighthouse, so called because of the dangerous eddies and currents that swirled around it, had withstood the fearsome gales blown in from the Atlantic since 1759. It had been built by John Smeaton, a man revered by Stevenson and considered to be the father of the civil engineering profession. Standing 70 feet high, it was made from interlocking solid portland stone and granite blocks, which presented a tall, smooth curved shape to the elements. It had been inspired, Smeaton said, by the trunk of an oak tree. "An oak tree is broad at its base," he explained, "curves inward at its waist and becomes narrower towards the top. We seldom hear of a mature oak tree being uprooted."

There had been several attempts at erecting lighthouses on the Eddystone rocks before Smeaton's triumphant endeavor, the most notable being the Winstanley lighthouse, built in 1698. Henry Winstanley, the clerk of works at Audley End in Essex, was also an enthusiastic inventor who took it upon himself to build a remarkable six-sided structure standing over 100 feet high on the Eddystone. With charming balconies, gilded staterooms, decorative wrought-iron work and casement windows for fishing, the whole curious structure was topped with an octagonal cupola complete with flags, more wrought iron, and a weather vane. It might have been more appropriately placed as a folly on a grand estate, but Winstanley was confident it could withstand the most furious of storms.

He was so confident that he longed to be there in bad weather to observe the might of the sea. By chance he *was* there, on November 26, 1703. That night a bad storm blew in with horizontal rain, screaming winds, and waves 100 feet high. Winstanley certainly got his wish. But at some time in the night, the fury of the sea took both Winstanley and his pretty gilded lighthouse and tossed them both to a watery oblivion. In the morning, nothing remained but a few pieces of twisted wire.

On his return from his trip in September, Stevenson immediately set about redesigning his lighthouse along the lines of Smeaton's Eddystone. He, too, would build a solid tower that curved inwards, the walls narrowing with height and accommodating the keeper's rooms. It would have to be at least 20 feet taller than the Eddystone, which was built on a rock above sea level, unlike the Bell Rock, which at high tide was covered by 11 to 16 feet of water. And if it was to be taller, it would also have to be wider at the base, over 40 feet, with solid, interlocking granite stone that would ensure it was invulnerable, even in roaring seas. More than 2,500 tons of stone would be needed, and Stevenson calculated that the cost of such a lighthouse would be around £42,000.

He could foresee that this cost would be a major obstacle, for the annual income collected by the Northern Lighthouse Board from dues was a modest £4,386. He was right: the board thought the cost prohibitive. They also questioned Stevenson's ability to undertake such an immense and difficult project. They felt he was too young and untried for this great responsibility, and pointed out that he had in fact only built one lighthouse before, a small lighthouse at that, and on the mainland. The board made it clear that they intended to consult established men in the civil engineering profession, men with a body of work and solid reputation, such as John Rennie, who was building the London Docks.

But Stevenson was a man who stood four-square to an unfavorable wind. The sweet wine of optimism flowed in his veins in generous measure and he took the negative epistle from the board as an invitation to his buccaneering spirit to try again. Meanwhile, the commissioners of the Northern Lighthouse Board realized they would never generate the huge sum needed for a lighthouse on the Bell Rock alone. It would need an act of Parliament to allow them to borrow the required amount, which they would then repay from the shipping dues they collected.

The first bill was rejected in 1803, but the subject was far from forgotten. The board was still hopeful for some sort of light and made it known that they would give consideration to any sensible plan that was submitted. A Captain Brodie stepped forward, with his plan for a lighthouse on four pillars made of a cast iron, plus a generous offer to provide, at his own expense, a temporary light until a permanent structure was in place. The board quietly shelved the plan for a lighthouse on cast-iron pillars but encouraged the temporary lights, which duly appeared, built of wood. And as each one was toppled by ruthless seas, it was replaced by Captain Brodie with growing impatience. Several other novice engineers also proposed plans for a lighthouse on pillars, including one who advocated hollow pillars, to be filled every tide by the sea. The conservative-minded members of the Northern Lighthouse Board remained unconvinced by them all.

The years were sliding by, during which Stevenson embarked on courses in mathematics and chemistry at Edinburgh University and worked on designs for other lighthouses. All the while, he was untiring in his efforts to interest the board in his now-perfected design for a strong stone tower based on the Smeaton plan. He envisaged a lighthouse standing over 100 feet tall, 42 feet wide at the base, with 2 feet embedded in the Bell Rock, and the whole exterior of the building encased in granite. The board members were polite but cautious. If only "it suited my finances to erect 10 feet or 15 feet of such a building before making any call upon the Board for money," Stevenson declared with growing impatience, "I should be able to convince them that there is not the difficulty which is at first sight imagined." While the officials procrastinated through 1804, a severe storm blew up and sank the gunship HMS *York* off the Bell Rock; 64 guns and 491 lives were lost. With the loss of a gunship at a time of Napoleon's unstoppable progress, the Admiralty at last woke up to the dangers of the Bell Rock.

Still the board took no action, and somewhat dejected, in December 1805, Stevenson could see no alternative but to send his plans to John Rennie, seeking his advice. Rennie was a man at the peak of his career, widely recognized as one of the best civil engineers in the country, with 20 years' achievement in building bridges, canals, and harbors. Nonetheless, Stevenson felt reluctant to share his ideas after all this time, pointing out that the design had "cost me much, very much, trouble and consideration." Rennie, fortunately, was greatly impressed by Stevenson's work and replied by return of post. He confirmed that only a stone building would survive the conditions of the Bell Rock and approved Stevenson's basic plan. He even came to a similar conclusion on the cost of the enterprise.

Rennie's approval was enough to unlock the door. Overnight, the Northern Lighthouse Board was transformed and unanimous: the commissioners wanted a lighthouse on the Bell Rock such as Rennie advocated. But first there would have to be a bill passed by Parliament allowing the board to borrow £25,000. In April 1806, Rennie and Stevenson went in person to Westminster to explain their case to the Lords of the Treasury and the Lords of the Admiralty. Progress was slow, but eventually the bill was passed and a date set for work to start on the infamous Bell Rock.

The Northern Lighthouse Board, mesmerized by Rennie's reputation and charmed by his charisma, placed him in charge, with Stevenson merely acting as his assistant. Rennie, himself, who had never built a lighthouse, argued that the light on Bell Rock should be a fairly faithful copy of Smeaton's Eddystone lighthouse, which was, after all, a proven success. He based the design of the lighthouse tower on this concept, but with a much greater curvature at the base to deflect the force of the waves upwards. Although a lot younger, Stevenson considered himself far more knowledgeable about lighthouses than Rennie, having now built several, and he was also familiar with the unique conditions of the Bell Rock. So without ever questioning the older man's authority, he became quietly determined to work entirely from his own plan.

But no one had ever built a lighthouse where so much of it was underwater. Stevenson could not know for sure whether a stone lighthouse was feasible. There was an awful possibility that the critics were right. Perhaps he was attempting the impossible, endangering life, and squandering money. The Bell Rock could so easily have the last word if his imagined sea citadel came tumbling down.

IN SPITE OF any misgivings, and after the years of delay, Stevenson was anxious to start work on the rock no later than May 1807. His first challenge was how to manage his workforce so far out at sea. No one had ever tried to build a lighthouse on a rock only exposed for 2 hours in every 12. Moreover, because the tide was later by about an hour every day, there would be times when the rock was uncovered only during the night. When building the Eddystone off the Cornish coast, Smeaton had ferried his men to work on a daily basis, but Stevenson did not see that as a practical proposition for the Bell Rock. He decided to take the bold step of keeping his men out at sea, at first on a vessel moored a safe distance from the rock and, later, in temporary quarters on the rock itself.

Stevenson hoped to erect a sturdy wooden building, known as the beacon, on the Bell Rock that would stand on timber beams well above the reach of high tide. At night, these temporary barracks would house the men and provide a warning light for passing ships. He was only too aware that such a building might be considered precarious, just a few feet above the swirling waters of the North Sea with no land in sight. He had not forgotten that Captain's Brodie's beacons had not withstood the relentless onslaught of the waves for long, nor that Winstanley's more substantial lighthouse had been blown away in the night like gossamer in the wind; but there was really no alternative. Should a sudden gale blow up, it might be impossible to row back in heavy seas to their vessel; and once the tide turned, the exposed rock would all too quickly be submerged again by the rush of incoming water. Some kind of temporary dwelling was essential.

Before the beacon could be built, he needed a ship that would fulfill a dual role: provide a dormitory for men working at the rock and display a floating light to warn ships at sea. The board was obliged to provide a warning light at night while work was in progress, which would enable them to charge dues from passing ships and to start repaying the loan. For this purpose Stevenson acquired an 82-ton vessel, the *Pharos*, named after the first celebrated beacon tower of ancient Egypt, the Pharos of Alexandria. The *Pharos* was then fitted out to provide 30 bunks for the workmen, quarters for the crew, and a cabin for Stevenson that would give him some privacy.

The *Pharos* would have to be moored one mile from the rock, since it was inadvisable to be too near the escarpment should she break anchor in bad weather. Here the water was particularly deep; and because high winds could easily set the ship adrift, a special heavy mushroom-shaped anchor was cast that would dig into the seabed and act as a drag. The men would have to row each day in small boats the mile from the *Pharos* to their work on the Bell Rock.

The starting date in May passed, and with it went the good weather. Stevenson was becoming impatient as preparations took longer than anticipated. Another ship had to be built, a 40-ton ship called the *Smeaton*, that would be used to bring out supplies to the rock. Stone from the quarries was ordered and masons were hired to cut each stone into its own individual design so that it interlocked with its neighbor and gave the tower stability. There were tools to be ordered; coal for the smith; food, alcohol, and water; and then the men to be hired who would labor possibly for years on the rock. Stevenson preferred to hire those who had worked for him before or had been recommended. He was a good judge of character and men usually stayed with him on wages of 20 shillings a week, "summer or winter, wet or dry," with rations of a half-pound of beef, 1 pound of bread, 2 ounces of butter, and 3 quarts of beer a day. The men also received the added bonus of papers, which protected them from the press gangs, who were quite ruthless in claiming men for service in the navy.

Stevenson made it quite clear to his men that nights would be spent onboard ship, and that no man could return home to his family for a month. After a month, he reasoned, the men would have adjusted to seasickness and, hopefully, fear, too. He worried that if they were allowed to go home earlier, they might not return. He also required the men to work on Sunday. This proved a problem to many, so before they began the epic journey to their new home in the middle of the sea—perhaps as some sort of insurance against the unknown dangers—the men crowded into the little church at the port of Arbroath to hear prayers.

Late in the evening on August 17, 1807, the *Smeaton* finally set sail. Ships in the harbor were flying their colors, and friends and family had gathered on the quay to see the men leave. As the ship moved slowly out of the harbor towards the darkening sky, the sound of cheers rang out across the water and echoed around the town before being lost to the sound of the waves.

They reached the rock, hissing and frothy with surf, at dawn the next morning. There was an air of excitement at being in such a strange place. It was too early to start work while the tide was still pushing water over their feet, so Stevenson raised three cheers to the men and poured a ration of rum for each. By 6:00 A.M. the water had retreated and some of the workmen began drilling holes for the beams that would support the beacon. The smith, James Dove, who would soon be busy sharpening tools, found a sheltered corner near a rock pool while other men cleared seaweed away from the pitted and uneven surface of the slippery rock. A seaweed called dulse was collected with enthusiasm; many of the men were suffering from seasickness and this was thought to be an antidote. When the tide returned, the men were thankful to row back to the relative security of their temporary accommodation on the *Smeaton*. As they pulled away, the rock that only minutes before had been a firm foothold was swallowed up before their eyes, with not even a ripple to mark its position.

Calm weather with whispering seas and wide, pearly-gold skies of late summer blessed the enterprise in the first few weeks. Stevenson's first task was to set up the forge. Everybody helped James Dove erect the iron framework, which would form the hearth. This was supported by four legs set up to 12 inches into the rock and secured with iron wedges. A huge block of timber, which would carry the anvil, was treated in the same way, and water was fast encroaching again as the weighty anvil was placed. Dove was invariably up to his knees in water, and sometimes up to his waist, but this was considered a minor problem compared to keeping the forge fire from the ever-playful waves.

The next task was to start work on the temporary hut, or beacon. This was uppermost in everyone's mind since, if there were an accident to the rowboats when attempting to land, this beacon on the Bell Rock would at least provide something to cling to until rescue arrived. Willing hands took on the difficult task of gouging out the hard sandstone that would take the stanchions supporting the uprights. Fifty-four holes in all, each 2 inches in diameter and 18 inches deep, were needed to hold the iron stanchions. The upper part of the stanchions aboveground would be riveted into the six massive 50-foot upright beams that formed the core framework of the beacon and other supporting beams.

One morning as the men rowed towards the rock, Stevenson was astonished to see what looked like a human figure lying on a ledge of rock. His mind was in turmoil, assuming that there must have been a shipwreck in the night and that the site would be littered with dead bodies. He was afraid his men would want to leave; they would see the Bell Rock living up to its reputation as a place of dread. As soon as he landed, and without a word, he made his way quickly to where the "body" lay only to discover, with immense relief, that it was in fact the smith's anvil and block.

Six days after leaving Arbroath, the men who had been very cramped on the *Smeaton* were transferred to the lightship *Pharos* now anchored a mile away. Everyone was pleased to be going to the larger ship, which had a well-equipped galley and bunks for the men. Her only drawback was that she rolled rather badly even in light winds. This made it extremely difficult for the men even to get into the rowboats for the mile-long row to the rock. Indeed, her rolling was so great, "that the gunwale, though about five feet above the surface of the water, dipped nearly into it upon one side," recorded Stevenson, "while her keel could not be far from the surface on the other." Everyone hoped the good weather would continue, not daring to imagine what she would be like if the weather turned. Seasickness, which had largely been conquered, now became a very big problem. Even Stevenson was affected.

On Saturday night, all hands were given a glass of rum and water, and every man made a contribution to the occasion—singing, playing a tune, or telling a story—so that the evening passed pleasurably, ending with the favorite toast of "wives and sweethearts." By Sunday morning, however, the atmosphere was much changed. There was the seriousness of breaking the Commandments to be considered. Several were opposed to working on the Sabbath, but Stevenson pointed out that their labor was an act of mercy and must continue without fail, although he emphasized that no one would be penalized for following his conscience. Prayers were said, and then Stevenson, without looking back, stepped into the boat. To his relief, he was followed by all but four of the masons.

Several days passed with work progressing well. The site for the lighthouse was marked out, a huge circle 42 feet in diameter in the middle of the reef, and the foundation holes for the beacon house were under way. On September 2, however, their luck changed. A strong wind blew up and a crew from the *Smeaton* who had rowed to the rock that morning, bringing eight workmen, were concerned that the *Smeaton* might break loose from her riding ropes and so took their rowboat back to check. No sooner had they reached the *Smeaton* than she broke from her moorings and began drifting at speed. The men who had remained on the rock were so intent on their work that they did not notice the rowboat leave or see that the *Smeaton* herself was floating quickly away.

Stevenson alone realized their terrible dilemma. He could see that with the wind and tide against her, the *Smeaton* could never get back to the Bell Rock before the tide overflowed the rock. There were 32 men working on the rock and only two boats, which in good weather might hold 12 men each. But now the wind was blowing in heavy seas. In such conditions, it would be fatal to put more than eight men in each boat to row the mile back to the floating light. It meant that there was transport for only *half* the men.

He saw the ship, too far away to help, and the men still involved in their work. As he stood there, trying to make sense of this seemingly insolvable problem, the waves came in with a sudden fury, overwhelming the smith's fire, which was put out instantly with a protesting sizzle and hiss. Stevenson himself was now "in a state of suspense, with almost certain destruction at hand."

With the obscuring smoke from the dead fire gone and the sea rolling quickly over the rock, the workmen gathered their tools and moved to their respective boats to find, not the expected three boats, but only two.

Stevenson watched helplessly as the men silently summed up the situation, only too aware of the rock fast disappearing under the sea. They waited. "Not a word was uttered by anyone. All appeared to be silently calculating their numbers, and looking to each other with evident marks of perplexity."

A decision had to be made. Soon the rock would be under more than 12 feet of water; they would have to take their chances. Sixteen men could go in the boats; the rest would have to hang on somehow to the gunwales while they were rowed carefully back through the boisterous seas to the *Smeaton*, now three miles away. There was no point trying to row to the *Pharos*, although it was nearer, as she lay to windward. Those clinging to the rowboats would stand little chance. Which men would have a place in the boats?

Stevenson was about to issue orders, but found his mouth so dry he could not speak. He bent to a rock pool to moisten his lips with the salty water and, as he did, heard someone shout, "A boat! A boat!" Looking up he saw a ship approaching—fast. By sheer good luck, James Spink, in the Bell Rock pilot boat, had come out from Arbroath with mail and supplies. As he approached, Spink had seen the terrible dilemma of those on the rock and had come to the rescue. This episode left a deep impression on Stevenson. The picture of the men silently standing by the boats awaiting their fate made him acutely aware of his responsibility for their safety, and that on this occasion, only a stroke of luck had averted a terrible catastrophe.

On September 5, 1807, the tide receded late in the day, and as the sea was running a heavy swell, making the rowboats hard to handle, Stevenson decided to cancel the trip to the rock. This proved to be a most fortunate choice, for the stiff breeze turned rapidly into a hard gale and would have made rowing back from the rock in the darkness a terrifying, if not fatal, experience. The storm raged all night and the next day; the little light ship was hit by successive waves of such force that for a few seconds, as she met each wave, her rolling and pitching motion stopped, and it felt as though she had broken adrift or was sinking. The skylight in Stevenson's cabin near the helm was broken and water poured in from the waves that were crashing on deck. Later in the morning, Stevenson tried to dress but was so violently thrown around in his cabin, he gave up.

At 2:00 in the afternoon, an enormous wave struck the ship with such terrifying force that tons of water poured into the berths below, drenching bedding and sloshing as one body from side to side as the ship moved. "There was not an individual onboard who did not think, at the moment, that the vessel had foundered," wrote Stevenson, "and was in the act of sinking." The fire had been extinguished in the galley, and the workmen, in darkness, were deep in prayer, swearing that should they survive, they would never go to sea again. But by evening the storm had blown itself out and the workmen were grateful to find the crew returning the ship to normal, with a fire in the galley and bedding being dried.

By mid-September, the bad storms that had prevented work on the rock were replaced by quiet seas and kind weather, during which Stevenson hoped to raise the six main beams of the beacon house. It was essential to get the 50-foot beams up and secure quickly, as a day or two of bad weather could destroy any work left unfinished, and autumn was approaching. More men were recruited from Arbroath. There were now as many as 40 carpenters, smiths, and masons on the Bell Rock. Their first task was to get a 30-foot mast erected to use as a derrick and a winch machine bolted down. The six principal beams—with iron bars and bolts already in place—were rowed on two rafts from Arbroath. In order to get the first four timbers securely in place in the space of one tide, the men worked in teams. They began before the tide was out, laboring deep in water, hoisting the beams into their allotted places. Others bolted them to the iron stanchions already fixed in the rock to a depth of almost 20 inches; still more men were ready to secure the uprights with wedges made of oak, then finally iron.

Every man worked with great intensity before the inevitable returning water reclaimed the rock. As the waves engulfed them, first up to their knees and then their waists, the four main beams were put in place and securely tied to form a cone shape; the timbers were temporarily lashed together with rope at the apex, and mortised into a large piece of beechwood. As the last men were leaving, up to their armpits in water, a rousing "three cheers" rang out from the men already in the rowboats. The beacon was standing bravely above the waves.

The next day the remaining two beams were put in place. Diagonal support beams and bracing chains were added, and the beacon was soon strong enough to hold a temporary platform where James Dove made himself a forge. At night, his glowing fire and shooting sparks presented an extraordinary sight as he perched above the waves. To everyone's great delight, the cumbersome bellows no longer had to travel back and forth to the rock with the rowers. The temporary platform was large enough to accommodate carpenters and masons who could work at making the beacon more secure even when the tide covered the rock.

Early October, with its shorter days and heavy seas, brought signs of the coming winter. Stevenson was pleased with the first season's work; the beacon looked capable of defying the winter, and a start had been made on the excavation of the lighthouse base. John Rennie paid his first visit, impressed with the progress, but expressing grave doubts about the durability of the beacon. His one night on the yacht, with the sound of the sea lapping so near his pillow was enough, and Rennie took his leave. By October 6, 1807, everyone departed, leaving the Bell Rock and its new beacon to face winter alone. As they made for land after almost two months at sea, the men watched the improbable beacon, sporting a cheerful flag, grow smaller on the horizon until, at last, it was swallowed up by the ocean.

WINTER AT HOME in Edinburgh could not have been more of a contrast for Robert Stevenson. Here he swapped the strictly male world of the rock, where a hard life—and possibly heroism, too—was traded for a daily wage, for a quiet family life with his wife and five children. In 1799, he had married Jeannie, Thomas Smith's eldest daughter by his first wife, and they lived with her parents in a large home Thomas Smith had built in Baxter's Place, Edinburgh.

Stevenson was becoming a man of some standing in the town as word of his involvement with the Bell Rock spread. His work did not entirely stop that winter: he undertook a small tour around the lighthouses for the Northern Lighthouse Board and his advice was often sought when problems arose in Thomas Smith's lamp-making business. The firm had won the contract for new street lighting in Edinburgh, and lights were also being custom-made in the workshop for lighthouses, which used many faceted mirrors as reflectors.

There was also a never-ending correspondence with John Rennie throughout that winter. As the chief engineer, Rennie was in charge of overseeing the progress on the Bell Rock but he was extremely busy managing large projects of his own. His work took him all over England and left him little time to visit Scotland. Despite this, he determined the shape of the tower, with a curvature rising gently from the rock at about 40 degrees in order to reduce the wave-force effect. He also insisted on dovetailing all courses of stone up to about 45 feet to provide greater strength. After this, he was content to send letters of advice, and Stevenson, sure that his understanding of the Bell Rock was superior, quietly ignored them. Rennie remained unaware of this, for Stevenson answered Rennie's letters with voluminous replies, full of complex questions on the lighthouse, designed to create a smokescreen of ambiguity.

As a result, there was never an actual schism, never an open disagreement between the two men. Stevenson always acknowledged the older man's mastery of current technology, but slowly, as work progressed, the balance of power shifted. Before long, Stevenson was openly working from his own plan for completing the lighthouse, sanctioned finally by the Northern Lighthouse Board.

Christmas passed, with warmth and fun centered on the children, but the cold winds of the New Year arrived, bringing with them a desolation for which the family was unprepared. The children succumbed first to measles and then to whooping cough, nursed patiently by Jeannie. But the illnesses proved too damaging for the 5-year-old twins, James and Mary, who died early in January. Their older sister, 6-year-old Janet, also weakened by illness, died just two weeks later. The Stevensons were devastated. They had seen their family decimated; Jeannie was inconsolable in the now-silent house. "Never was there such a massacre of the innocents," Robert Louis Stevenson wrote years later. "Teething and chin cough and scarlet fever and small pox ran the round; and little Lillies, and Smiths and Stevensons fell like moths about a candle."

The grief of the winter left Robert Stevenson ready for a change of scene, to be active again in the company of men, taking on tangible difficulties that could be understood and overcome. The dominant thought in his mind was the lighthouse and, above all, the beacon. Rennie had strongly advised against it, fully expecting it to disappear over the winter. Stevenson was more optimistic, but he had to know for sure.

Taking advantage of a spell of calm weather at the end of March 1808, Stevenson set out in the lighthouse yacht with some of the workmen on the evening tide. As they approached the Bell Rock in the early morning light, he strained for any sight on the horizon. There it was: the beacon, riding the waves, just as they left it last October. "There was not the least appearance of working or shifting at any of the joints or places of connection," Stevenson wrote triumphantly, "except for the loosening of the bracing chains, everything was found in the same entire state in which it had been left." He was now sure enough of the strength of the beacon to build it up further and use it as a refuge should the need arise. If all went well, he hoped at last to make quarters for the men above the tide.

The season's work began in earnest when Stevenson, with his foreman, Peter Logan, and a dozen masons set out in a new custom-built schooner, the *Sir Joseph Banks*, on the afternoon tide of May 25, 1808. His priority over the coming months was to dig out the foundation pit for the lighthouse in the middle of the Bell Rock and create a small cast-iron railway, some 300 feet in length, that would run from the various landing stages to the site of the lighthouse. Strange though it seemed to build a railway out at sea, Stevenson decided it was the best way to transport the one-ton blocks for the lighthouse across the rock with its corrugated and spiky surface. Any damage done to the individually cut blocks, each one exactly shaped to interlock with its neighbor like a three-dimensional jigsaw puzzle, could render it completely useless.

With good weather continuing through June, James Dove and his assistant at his forge on a platform on the beacon were endlessly engaged making the fittings for the railway and sharpening tools. Meanwhile, masons labored at the rock surface, shaping the 42-foot-wide circle for the lighthouse foundations. Stevenson had decided against using explosives to excavate the 2-foot foundation into the solid rock in case irreparable damage was done. Therefore, every piece of rock had to be laboriously dug out by pickax. The deeper they excavated, the harder the rock became. One man was continually employed replacing pickax handles while others were called to help bail out the foundations after each tide.

Meanwhile, the landing master and his crew of sailors helped the millwrights to lay the cast-iron railway. They built up the surface where necessary, sometimes as much as five feet to accommodate the jagged unevenness of the rock. Viewed from the sea, the Bell Rock presented a strange sight. There were sometimes as many as 60 men crowded on its surface, "the two forges flaming, one above the other," wrote Stevenson, "while the anvils thundered." The entire work party was surrounded by clouds of hissing steam, oblivious of the crashing waves below. The sight was even more strange by night. If it was low tide, the men continued working by light of the forges, strange dark shapes with their torches casting unnatural shadows over the pitted rock surface.

The calm seas and warm summer weather enabled the men to gouge out the reluctant rock in good time and by mid-July the 2-foot-deep foundation for the lighthouse was ready. Irregularities in the floor of the rock had to be leveled with 18 specially cut stones. The largest of these, the massive foundation stone of some 20 cubic feet, was engraved with the year 1808 and fitted with due ceremony. Flags were fluttering from the boats and the whole workforce was in a celebratory mood. They stood in the sunshine and cheered as Stevenson and his chief assistants, Peter Logan and Francis Watt, applied the token pressure from level and mallet. "May the great architect of the universe complete and bless this building," said Stevenson, in a moment of prayer. The motley workforce standing, it seemed, almost on water, were absolutely confident that under Stevenson's leadership they could build this sheer and slender column that would defy the power of sea, wind, and the slowly creeping rust of time. The whole surface was smooth and ready and the construction of the lighthouse could now begin.

As the work gathered momentum, many workers chose to sleep on the beacon rather than row all the way back to the bunks in the *Sir Joseph Banks*. Since the railway was still not finished, every block had to be manhandled across the difficult terrain, yet by mid-August, the first course of 123 interlocking blocks was in place—no small feat since the total weight of these blocks was 104 tons. A fortnight later, the second course was laid and treenailed into position. Oak was used for the treenails, which were slotted into previously drilled holes in the stone. These were then lined up with holes in the stone below to join one course to another. The men were much inspired now to see the bold curving lines of the lighthouse taking shape and were as keen as Stevenson to see the third course completed before work finished for the season.

Back at the yard in Arbroath, the engineers' clerk, Lachlan Kennedy, was conscious of the race against time, too. Kennedy arranged for the yard workers to load the *Smeaton* by torchlight during the night of September 20 with the last 17 stones of the third course, so that the ship might catch the 2:00 A.M. tide and be at the Bell Rock by breakfast. Stevenson and the men were delighted to see the *Smeaton* unexpectedly sail into view in spite of heavy seas and strong tides that morning, since it now looked possible to finish the third layer of blocks. Any satisfaction soon evaporated, however, as they became helpless witnesses to what followed.

Two men from the *Smeaton*—the mate, Thomas Macurich, and one of the crew, James Scott, a lad of 18—went in a small boat to attach the *Smeaton*'s hawser to the floating buoy. This had become caught on a rock on the seabed, shortening the chain and depressing the buoy; the attaching ring of the buoy was barely above water. While the two men struggled with heavy rope to secure the *Smeaton*, the chain suddenly became disentangled and the large buoy vaulted up with such force that it upset their little boat. The two men were thrown into the sea.

While the mate managed to cling to the gunwale of the boat, James Scott had been knocked unconscious and was unable to grasp anything that would save him. Before anyone could attempt a rescue, he was carried away by the strength of the current and he disappeared! A distress signal was immediately hoisted, but he was never found. It was an unsettling tragedy for the men on the Bell Rock to witness, particularly as Scott, with his pleasant and willing disposition, was everyone's favorite.

It was agonizing news to have to break to Scott's family. His mother was overcome with grief. To add to her difficulties, her seaman husband was being held in a French prison and James was the only one who had been bringing home a wage to her and her four other children. It occurred to the men that to alleviate her hardship, James's younger brother might take his place on the *Smeaton*, but no one could find the courage to suggest this to her. Eventually, the landing master, Captain Wilson, seeing her desperate circumstances, put the idea forward. "Such was the resignation, and at the same time, the spirit of the poor woman," wrote Stevenson, "that she readily accepted the proposal and in a few days, the younger Scott was actually afloat in the place of his brother."

The third layer of stones was duly laid, making the lighthouse four feet high, although the achievement was robbed of its glory by the death of James Scott. Late in the season, with winds growing stronger, everything on the rock was made safe and left to the mercy of the winter seas. The busy days of summer were soon just a memory as the emerging lighthouse, the railway, and the marks of humanity were claimed by the ghostly underwater world of winter at the rock.

STEVENSON MADE HIS way back to the comforts of home, but not for long. He took the *Smeaton* on a short lighthouse tour in terrible weather, during which the ship was pounded within an inch of destruction and Rennie made his annual visit, braving the ongoing storms, and was shown the season's work on the rock while suffering his usual agony of seasickness. That he approved of Stevenson's work was now clear to the solid, conservatively minded members of the Northern Lighthouse Board, who found they had in their employ, by some whim of fortune, Robert Stevenson, the coming man.

Stevenson's greatest concern during the second winter was to find and prepare the best granite for the body of the lighthouse. This was not easy, as most quarries could not produce enough rock of the size required. His search took him on long journeys all over Scotland. He favored granite from the Rubislaw quarry at Aberdeen and transported it back in huge, uncut six-ton blocks. Several more craft were needed for the complex operation of transporting all the stone to the yard at Arbroath, where it was cut and prepared before being taken out to the rock. Two more ships were added to the fleet, the *Patriot* and the *Alexander*, as well as wide, flat-bottomed boats called praams, which would take the finished stone from the ships to the Bell Rock itself. One final crucial link in this chain of transport was a horse named Bassey. Unaided, this worthy animal carried the whole of the lighthouse, 2,835 stones, block by block from the yard to the quay.

Despite Stevenson's determination to make good progress at the rock, the harsh winter weather lingered into spring; it was snowing when the third season's work started in May 1809. The men began by repairing winter damage to the railway and continuing its progress around the lighthouse. Joists to create another floor level, and battens for walls were added to the makeshift beacon as well. In spite of its frighteningly precarious appearance, frequently now the men competed to determine who would spend the nights there in preference to a berth on the *Sir Joseph Banks*. The beacon had now survived two bleak winters and confidence in it was growing.

This confidence was rudely shaken within a few days though, as shrill winds from the cold wastes of the north carried with them a storm to remember, with mountainous seas. Stevenson, onboard the *Sir Joseph Banks*, was hardened to danger after years of sailing around the Scottish coasts, but even for him, the "rolling and pitching motion of the ship was excessive . . . nothing was heard but the hissing of winds and the creaking of bulkheads." His greatest fear was for the 11 men who had chosen to stay in the unfinished beacon. For 30 hours, in a building with no proper roof, no food, bedding, or comfort of any kind, these men were exposed to the terrifying storm. As soon as was possible, the crew of the *Sir Joseph Banks* set out for the rock with supplies and a kettle full of mulled wine, wondering what they would find.

They were greeted by a scene of devastation. Huge seas had washed everything in the house away, even some of the lower floor. The beacon had suffered "an *ill-fared twist* when the sea broke upon it," Peter Logan said with some understatement. But miraculously, all the men had survived. It soon emerged that during the storm their spirits had been kept high by a joiner named James Glen, who had regaled them with stories from his early adventures as a sailor "after the manner of the *Arabian Nights*." Glen had recounted such horrific tales of hardship at sea that the men were inclined to think that their own fate, in what amounted to a makeshift box perched just above the ravenous waves, was preferable. Indeed, so inured were they to the petty trials of life that Watt and Glen continued to sleep at the beacon house, declaring that they "were not to be moved by trifles."

Although conditions on the rock were unpleasant, buffeted by strong winds and heavy seas, the beacon house was soon repaired and the space separated into individual rooms. On the lowest floor, the smith was at work, sharing the space with the workers mixing mortar. The second floor provided sleeping quarters, with bunks for those who preferred to sleep at the beacon. A galley and a small cabin for Stevenson and Logan were added. Balking against unpleasant conditions, Stevenson would not allow anything to slow progress on the lighthouse. He need not have worried. By June, the tower was high enough for a rope bridge to be connected from the beacon house to the lighthouse, and work continued as quickly as it could, regardless of the tide. In fact, Stevenson almost had to restrain the enthusiasm of the men who would often row to the rock at daybreak, at 4:00 in the morning, to start their day's work.

By the end of June, the tower had risen to a height of 12 feet. The biggest problem was that the mortar was being continually washed away by heavy seas, although an especially hard, quick-drying Pozzolano mortar was being used. There was no easy solution to this problem, except to hire more mortar mixers who were always prepared with fresh supplies.

Another problem was more easily solved: At this level the tower was too high to use the derrick cranes, so an ingenious new balance crane was adapted by Watt especially for the conditions of the tower. From a central column, a horizontal jib could be raised or lowered, the weight of the article lifted counterbalanced by a similar weight at the other end of the jib.

By June 30, it was necessary to move the crane higher. It was now 35 feet above the rock and quite difficult to maneuver. At one point during the operation, someone neglected to tie a purchase rope securely, which resulted in the crane and a ton-weight on the jib falling with a terrible crash on the rock. Men ran in all directions—except for the principal builder, Michael Wishart, who had tripped directly under the crane. His feet were caught and badly mangled. It looked as though amputation would be the only answer. White with shock and from excessive bleeding, he was stretchered into a boat and hurried back to Arbroath. When Stevenson visited him a few days later, it was clear he had escaped amputation and was making a good recovery. He could no longer help to build the lighthouse, but he expressed the hope to Stevenson that he might "at least, be ultimately capable of keeping the light at Bell Rock."

By mid-July, high tide no longer covered the lighthouse. The beacon house was also complete, with a roof of tarred cloth and moss lining the space between the walls for insulation. The walls themselves were covered with green baize, which created a much more homey effect. Now most of the men wanted to spend their nights there, including the new cook, Peter Fortune. Fortune was a man of many talents, apart from working as cook and surgeon, for which he was paid 3 guineas annually; he also served as the barber, steward, and provisions accountant.

On July 22, 1809, an express boat arrived from Arbroath with the news that an embargo had been placed on all ships while the army was en route to the Netherlands. The lighthouse ships were forbidden to leave Arbroath and the *Smeaton* could not supply the blocks that were needed to complete the half-finished thirteenth course. Stevenson, fretting at the delay, hurried to Edinburgh, where he argued that the lighthouse was a special case. The customs officials, however, were not to be persuaded and referred his appeal to the Lords of the Treasury in London. The situation looked bleak until Stevenson managed to coax the customs men into letting him take the necessary blocks and provisions out to the rock in the company of a customs officer.

By early August, the tower was 23 feet high and the lower crane at ground level was now raised on a platform of blocks 6 feet high. This extra height enabled the lower crane to deliver blocks to the crane on top of the building. Work was proceeding well, although the hot, still days of summer were few and far between. The men at the rock were constantly harassed by angry seas, soaking their clothes, flinging their tools about, and washing the mortar from newly laid stones. By August 1, 50-foot waves were crashing down with great violence upon the beacon. Yet in spite of these repeated attacks, and with successive seas finding new weaknesses, there was a nucleus of men with courage beyond the everyday who would not give in to the elements.

On August 20, a complete course of 53 stones was laid in one day; and less than a week later, the solid part of the lighthouse, consisting of 26 courses, was finished. It stood more than 31 feet above the rock, 17 feet above high tide, and marked the end of work for the season. The last stone of this significant level was laid with full ceremony before the men set off to Arbroath with flags flying. But Stevenson was frustrated at ending the work in August. There was still another 70 feet to build before the light could shine out. The whole season of 1809 had been plagued by bad weather, then the embargo. Nonetheless, he judged it best to leave the solid stump of lighthouse to the mercy of the winter seas, rather than try to take the walls higher that year. Besides, he needed another crane since the one in use had guy ropes that were too long to be manageable in the face of the autumn gales.

Over the winter he hoped to appoint the lighthouse keepers. Stevenson decided that the Bell Rock must have at least four: three on duty and one on leave in Arbroath. He did not want a repeat of what had happened at the Eddystone, where there had only been two keepers on duty. When one man had died, the remaining keeper, fearful of being accused of murder, had not informed anyone; and when the lighthouse had been visited a month later, the body had been found in an advanced state of putrefaction with the surviving keeper decidedly unbalanced.

It did not take Stevenson long to decide on his first keeper. John Reid, captain of the *Pharos*, was, he thought, a person "possessed of the strictest notions of duty." As his assistant, he chose Peter Fortune, the cook, who had "one of the most happy and contented dispositions imaginable."

Stevenson wanted the light on the Bell Rock to be very bright and easy to identify so it would not be confused with any other lighthouse. To that end he set out on a tour to inspect other lights. This led to a number of technical improvements: the silvered surface of the reflectors, which directed the beams of light, had to be accessible for polishing to prevent the light from being dulled; and he devised a system to collect any oil that spilled below the burners. His key concern, however, was how to give the Bell Rock a unique signal.

Stevenson was particularly impressed with the new Flamborough Head lighthouse in England, which used colored light—two white beams followed by a red beam. The effect was created by mounting the reflectors on a rotating axle, which revolved around the light. The reflectors were arranged in three rows, with one of the rows covered in colored glass. Stevenson also realized the mechanical device for moving the reflectors could control the striking of a bell, to give extra warning in heavy fog. Back in Scotland, he set to work experimenting with different designs and colors. Red invariably gave the strongest light. The Bell Rock lighthouse, he decided, would flash a bright white light and then, at closer range, a red flash would alternate with the white. It was a design that would set the standard in Scotland for years to come.

—

THE WINTER BREAK at least gave Stevenson a chance to be with Jeannie again. She was heavily pregnant over Christmas and in February 1810 gave birth to a baby boy they named James. Stevenson enjoyed barely two months at home with his new son before salty seaspray at the cold and unwelcoming Bell Rock came into view for the fourth season. It was mid-April and there were still 66 courses to build. Through these stages, the lighthouse would be narrower and no longer solid, so the stones would require very careful cutting if they were to fit into the exacting requirements of the spiral staircase, the separate rooms, the ceilings, and the windows.

The first essential job of the spring was to build a substantial wooden bridge connecting the lighthouse to the beacon. This would speed up work, as the heavy seas made it impossible to even set foot on the rock for almost a month. They soon also found that seaweed had festooned the lighthouse and made walking on top difficult. It was clear, too, that during their absence, the sea had broken into the beacon house, discovering the secrets of the rooms, visiting and revisiting, and leaving its watery mark on the green baize. A new balance crane was quickly erected, its various parts landed by the praam in a strong easterly wind and hauled into position. Still it was out of the question to start work in the blustery weather, which imprisoned the men in the beacon. The delay worried Stevenson; he dreaded the idea of the lighthouse left unfinished by late summer, overtaken by bad weather as the higher courses were being laid. It would be impossible to send men up as high as 80, 90, 100 feet to lay masonry in the autumn gales, when wind gusts of 90 miles per hour were not unusual.

Work started in earnest by May 18, and the men entered into a race against time regardless of the weather. By May 25, the building had reached the door lintel. Less than three weeks later they were covering the ceiling of the first room. By the end of June, the lighthouse assumed a fine inward curve and stood 64 feet high. Now the situation was getting dangerous, for there was less room to work and a long way to fall; inattention or carelessness could be fatal. The job of pointing the walls was the most precarious; the men "stood upon a scaffold suspended over the walls in a rather frightful manner," wrote Stevenson. As the tower increased in height, it was a constant fight to keep balance in the wind gusts, and many workers lost clothing, jackets, and hats to the sea below them.

Work was again hindered in early July when storm-force winds battered the rock. Heavy seas pushed up from below the beacon house, carrying everything away, including provisions and equipment from the lower floor. It took two days to repair the damaged upper course of masonry where the sea had gouged out the mortar. Whatever the elements did now, however, the end was in sight. By July 21, the last principal stone of the building was hauled to the quay in Arbroath by the indomitable horse Bassey, who was embellished with bows and streamers for the occasion. At the end of July, the last stone was laid to the ninetieth course by Stevenson. He was surrounded by azure skies and placid seas, and every boat proclaimed the victory, hoisting colors and flying flags, the men cheering. The lighthouse now soared 102 feet 6 inches. "May the great architect of the universe, under whose blessing this perilous work has prospered, preserve it as a guide to the mariner," Stevenson said with his customary ceremony as he laid the stone.

All that remained was to fit out the lightroom. But on the August 15, with the final preparations being made, a storm—the like of which no one had yet encountered—hit the rock with awesome fury. The *Smeaton* broke anchor and drifted directly for the Firth of Forth. The men "cooped up in the beacon were in a forlorn situation, with the sea not only raging under them, but also falling from a great height upon the roof." The bridge to the lighthouse was completely impassable as the sea gushed constantly over it. With 80-foot waves, the lighthouse itself, still without a roof, was without protection as the sea poured over the top of the walls and out of the windows and door. The storm lasted three days while the men on the floating light could only watch on in horror as, first, the lower floor and then most of the one above, was torn from the beacon by the waves. When the storm had washed itself away, the 17 men in the beacon, pale and exhausted but alive, started once again on repairs and clearing up.

By October 16, the glazing in the lightroom was almost finished. That evening, James Dove gave orders for work to cease, and the team of workers made their way down through the lighthouse and across the bridge to the beacon house for dinner. Two friends, Henry Dickson and Charles Henderson, were fooling around now that the day's work was over, and made a competition of getting there first. They tried to outrun each other as they descended the lighthouse, Henderson in the lead. When Dickson reached the cookhouse he was surprised not to see his friend, who had been ahead. No one in fact had seen him; and no one had seen him fall into the darkness of the water as he had lost his footing hurrying over the rope bridge. A search was made of the black and silent water, but he was never found. It was the second tragedy in two years, but the men, determined, worked on.

The lightroom was at last finished by the end of October. The reflectors were in place, the lamps made by Thomas Smith were fitted, the revolving clockwork machinery set up, and two 500-weight "fog bells" were placed on the balcony. All was ready except the defining sheets of red glass, 25 inches square. Red glass of this size was not made in England, and the one man in the country, James Oaky, who had been prepared to try, had not yet been inspired to create this important article. A foreman was sent to London with instructions to stand over Oaky until the red glass was produced. The brilliant red glass was soon complete.

On February 1, 1811, the blackness enveloping the terrible power of the Bell Rock was banished forever. A shooting ray of white light, followed by red, cut the night. Everything that Stevenson strived for, everything his team of men had struggled for at such great risk to life, evaporated in the dazzling beam of light, which sent out John Gedy's original warning from the fourteenth century. The lighthouse that everyone had said was impossible now stood on top of the impregnable and ancient rock, its ability to destroy conquered. Huge seas could hammer at the slender column, seaspray 100 feet high could hiss and curl around the light diffusing its brilliance, but it would stand a glittering star in the wildest storms.

It was an enormous achievement, and brought Stevenson wide recognition and acclaim. He had been young and relatively untried when he built the Bell Rock lighthouse, but on the strength of this success he set himself up in business as a civil engineer, building lighthouses, bridges, harbors, and roads. The Northern Lighthouse Board continued to employ him; he built 18 lighthouses in all, often astride an impossible precipice whipped by the sea on some bleak windblown rock. His sons and grandsons followed after him, festooning the rocky shores all around Scotland with sparkling beacons in the night.

He and Jeannie lived on in their big house, so loved by their children and grandchildren, with its mysterious cellars and secret lofts full of apples ripening and a lifetime's treasures. They had been married for 46 years, their lives entwined more than they knew. So when Jeannie died in 1846, he found her absence unbearable. His own life had overflowed with riches exactly to his liking—full of comradeship and adventurous trips. His grandson, Robert Louis Stevenson, later wrote about the "intrepid old man," who, when told he was dying, fretted not at approaching death, which he had faced many a time, but at the knowledge that "he had looked his last on Sumburgh, and the wild crags of Skye, and the Sound of Mull . . . that he was never again to hear the surf break on Clash-

Isambard Kingdom Brunel in 1857,
shortly before the first attempt to launch
the *Great Eastern* into the Thames.

Brunel's early sketches for his "Great Ship," from the Brunel Sketch Books.

Preparations
to launch the
Great Eastern,
1857–58.

Isambard Kingdom Brunel in September 1859, aboard his "Great Ship" as she was being prepared for her maiden voyage. Within a few hours of this photograph being taken, he collapsed with a stroke.

Isambard Kingdom Brunel did not live to see his "Great Ship" on her maiden voyage to America, in 1860, with just 38 passengers aboard.

Robert Stevenson in 1814. He won much acclaim for building the Bell Rock Lighthouse 11 miles out to sea.

Designs for the Bell Rock Lighthouse and the beacon, which was to house the men on the rock during construction.

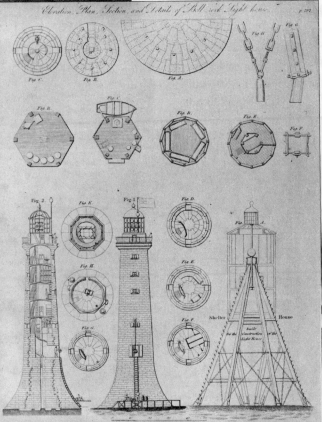

The Bell Rock Lighthouse nearing completion in 1810.

The Bell Rock Lighthouse during a storm at sea,
engraved from a drawing by J. M. W. Turner.

John Roebling, designer
of the Brooklyn Bridge,
photographed a few years
before his death in 1869.

Washington Roebling,
his oldest son, who carried
out his father's vision.

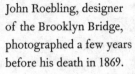

Emily Warren Roebling,
Washington's wife, who helped her
husband complete the bridge when
he became too ill to leave his room.

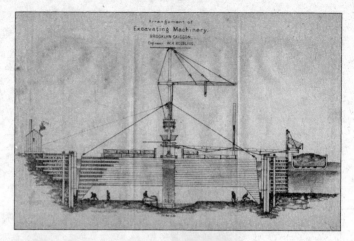

Designing the foundations for the Brooklyn Bridge under the East River. A cross section of the Brooklyn caisson drawn in 1869.

Cable wrapping high above the East River in 1878. There were thousands of wires to a cable, which were then tightly bound with an outer skin of more wire.

The Brooklyn Bridge under construction. Once the towers were complete, the first cables could be spun across the river.

When the main cables were in place, suspenders were
attached that would hold the steel floor beams.

The Brooklyn Bridge circa 1890. When the bridge was completed, in 1883,
it dominated the New York landscape and was acclaimed as a wonder of the world.

carnock; never again to see lighthouse after lighthouse open in the hour of dusk their flowers of fire, or the topaz and ruby interchange on the summit of the Bell Rock."

He died in July 1850 at the age of 78. There were, of course, many tributes praising a life so full of achievement, including one from the Northern Lighthouse Board, whose commissioners set a marble statue in the Bell Rock lighthouse with the words: "in testimony of the sense entertained by the Commissioners of his distinguished talent and indefatigable zeal in the erection of the Bell Rock lighthouse." To this day it shines out over the North Sea and remains the oldest offshore lighthouse still standing anywhere in the world.

CHAPTER 3

THE BROOKLYN BRIDGE

When constructed in accordance with my designs
[the Brooklyn Bridge] will not only be the greatest bridge
in existence, but it will be the great engineering work
of this Continent and of the Age.

JOHN AUGUSTUS ROEBLING, *1867*

IN THE EARLY NINETEENTH CENTURY, the great East River was seen as an unbridgeable gulf between two fast-growing American cities: New York on Manhattan on its western bank and Brooklyn on Long Island on its eastern side. The deceptively named East River is actually a tidal strait, an arm of the sea a third of a mile wide that reaches around Long Island separating it from Manhattan. Its tides are fast and furious, made worse by whirlpools and perilous currents caused by hazardous peaks and troughs on the floor of the river. Ships were often compelled to wait in the bay for an incoming tide before they could make headway inland safely.

As the population increased during the nineteenth century, the once-sleepy villages surrounding Brooklyn became rapidly engulfed, their rural identities forgotten. By 1850, the bustling town of Brooklyn vied with New York as a center of commercial enterprise. The major impediment to a vibrant growing trade and the further advancement of both cities was the East River itself, cutting like a knife through the budding prosperity. There was a ferry service, but this was not equal to demand, and in the extreme cold of winter often ceased altogether. When the river froze, those who chanced the journey on foot could find themselves suddenly in extreme danger as the swiftly returning tide broke up the ice. "The only thing to be thought of is a bridge," declared the *New York Tribune* in 1849, giving voice to popular opinion. But who could build a bridge across such wide and turbulent waters?

During the winter of 1852, the river once again turned to ice, the ferries often trapped, their dark shapes looking doomed, imprisoned in the icy expanse. For one trapped ferry passenger, John Augustus Roebling, traveling with his 15-year-old son, Washington, an idea began to take shape during those enforced hours spent in the freezing no-man's land of ice, with both cities so temptingly near and yet so impossibly out of reach. The river was deep, the currents and tides overwhelmingly difficult, the distance enormous, but John Roebling did not willingly bend to the inconvenience of what was considered impossible. He felt sure he could overcome these obstacles and build a bridge of strength and beauty such as the world had not yet seen.

Roebling had come to the New World from Europe to make his mark, and everything he did carried his signature of originality. He was a man cast in an heroic mold, with the look of some visionary Old Testament prophet; he had an iron will and sense of purpose and eyes that saw more than most. As an engineer, he had built bridges before, flinging causeways over preposterous gorges and supposedly unbridgeable rivers, turning granite and steel and blood and sweat into practical works of art. His bridges at Pittsburgh and Cincinnati were much admired; trains ran over the frothy Niagara on his suspension bridge. And as he waited on the ferry and looked across the ice at the two cities divided, his mind was on fire with a new idea. His brilliant scheme would come to be admired the world over, a technological revolution in bridge design celebrated both for its incredible strength and beauty—but it would also lead to the ruin of both its creator and his young son. "A bridge always claims a life," said industrial folklore and superstition. Roebling's "Great Bridge" at Brooklyn would demand an even greater sacrifice.

Back in his hometown of Trenton, New Jersey, Roebling immediately set to work on his bold plan for the largest suspension bridge yet built in the world. In 1826, the acclaimed suspension bridge built by Thomas Telford over the Menai Straits in Wales had set a record with a span of 579 feet. The Wheeling Bridge of West Virginia, created in 1849, set a new record with a span of over 1,000 feet. Roebling, however, was contemplating a bridge over the East River with a heretofore unthinkable span: from tower to tower of 1,600 feet. At either end he envisaged massive gothic towers, like some medieval cathedral, rising sheer from sea level to nearly 300 feet, dwarfing the world below. The very essence of strength and durability, the towers would bear the weight of a roadway nearly 80 feet wide and hold in tension the lace of steel wire as it soared from the road to the tops of the towers and down again on the other side.

Despite Roebling's great reputation, many leading figures in Brooklyn and New York were skeptical; it seemed just a hopeful dream. Suspension bridges had a history of spectacular failure. The decking of these bridges was notoriously unstable and prone to roll; the wind alone could turn the most stable bridge into confetti. Telford's admired bridge over the Menai Straits had been damaged by a storm in 1839. Five hundred soldiers crossing a suspension bridge at Angers, in France, had been drowned in the river below when the bridge crumbled beneath their step. Even the famous Wheeling Bridge in Virginia had not lasted 10 years.

Time passed, the American Civil War was fought and won, and Roebling's bold proposal went quietly ignored even though his reputation as an engineer could not be challenged. A German immigrant, born in 1806, in the sleepy town of Muhlhausen in Saxony, Roebling had studied civil engineering and received his degree in Berlin in 1826. In his early years in America, he had worked as a surveyor and invented a strong wire rope made of iron, to replace hemp, which he made in a shed behind his house in Trenton. In 1845, he had built his first suspension bridge in Pittsburgh, over the Allegheny River. This had been followed by his spectacular railway suspension bridge over Niagara, which was such a novelty that one nervous passenger, Mark Twain, observed: "You divide your misery between the chances of smashing down two hundred feet into the river below, and the chances of having a railway-train overhead smash down onto you." Twain need not have worried; Roebling's great experience in the art of bridge building had taught him how to incorporate strength and stability into the design. Over the years, he developed new systems for securing the main cables and stiffening the bridge floor, and he introduced iron stays, radiating from the towers to the floor of the bridge.

It was not until the freezing winter of 1866, while Roebling was working on his much praised Cincinnati Bridge, that the East River again became blocked with ice, for weeks. At last, demands for a bridge could no longer be ignored. Brooklyn had become America's fastest-growing city, with 400,000 inhabitants; more than a thousand ferry crossings were made each day. A well-established contractor based in Brooklyn, William Kingsley, who was keen to promote the idea of a bridge, became convinced Roebling's plan would work. He persuaded Henry Murphy, an influential senator and Brooklyn lawyer, that the enterprise was possible, and soon Roebling's proposal began to gain widespread support.

In January 1867, a bill was passed creating the New York Bridge Company, which was to build the Brooklyn Bridge from private funds, and Roebling was soon appointed as chief engineer. A meticulous and hardworking man, within three months of his appointment he presented the New York Bridge Company with detailed plans for his Great Bridge. Although this was to be the biggest suspension bridge in the world—

more than twice the span of the Clifton Bridge and nearly three times that of the Menai Suspension Bridge—Roebling assured the Bridge Company that its enormous scale need not compromise its safety. "A span of 1,600 feet or more can be made just as safe, and strong," he declared, "as a span of 100 feet."

The two mighty towers, over 276 feet high and of gothic design, would dominate the scene. "These great towers would serve as landmarks to the adjoining cities," he declared, "and will be entitled to be ranked as national monuments." They would stand on massive foundations that would reach bedrock, around 40 feet below water level on the Brooklyn side and more than 70 feet below water level in the deeper water of the New York side. Two enormous anchorages would hold the suspension cables on either side of the river. Each of the four cables would then enter the anchor walls at a height of nearly 80 feet above the tide and connect to anchor chains, which in turn would be attached to massive anchor plates. From the tops of the towers a system of stays radiating diagonally down to the floor would confer additional strength to the roadway. And to ensure "the degree of stiffness essential to meet the effect of violent gales," Roebling proposed six iron trusses to run the whole length of the suspended floor from anchor wall to anchor wall. Finally, he proposed that the cables should be of steel, not iron. This alone was a revolutionary new idea; steel had never been used on such a vast project.

The gentlemen of the New York Bridge Company were impressed with Roebling's report, with his thoroughness, his experience, his attention to detail, and his patient explanations. With the towers soaring higher than any other building in New York and a total estimated length of the whole bridge almost 6,000 feet, it became clear to the company that the bridge would be a wonder of the modern world. Almost by some wizardry, it seemed, he had placed before them a design of such grandeur that their response was unanimous; their confidence in the man and his ideas was total. "When constructed according to my designs," the charismatic Roebling concluded, "it will not only be the greatest bridge in existence, but it will be the great engineering work of the Continent and of the Age!"

Money for Roebling's bridge was raised by issuing stock; the city of Brooklyn bought $3 million, New York a smaller amount. The general public also bought stock in the bridge and eventually $5 million was raised. At last, in 1869, after years of gestation, the project could get under way. Roebling was 63, but still a man of strength who was fired with enthusiasm for the bridge. His whole life had been spent, his energies and his ambition directed, towards this goal.

Roebling's vision was shared by his son Washington, who had been with him in 1852 when they were trapped in the ice on the East River. Together they would create this masterpiece. But on June 28, 1869, while engrossed with his son in routine surveys for the exact site of the Brooklyn tower, John Roebling failed to notice a ferry coming too fast. It crashed into the Fulton slipway where he was working.

In the accident, Roebling's foot was badly crushed; he was taken to his son's home nearby, where it was soon clear that his horribly mangled toes would have to be removed. He refused to take any opiates; his mind was not to be clouded, so the operation was performed without anesthetic. To the mounting concern of his family, he also directed his own treatment, dismissing his doctors. Firmly believing in the curative power of water, he insisted that water be trained on his raw wound day and night. Then, while he fretted at being laid up when there was so much work to do, slowly, insidiously, lockjaw set in.

In a matter of days the muscles in his face became rigid, unmovable, set in a frightening death mask grin, teeth locked tight together. He could not speak. He could not eat. Severe pain and convulsions followed, contorting his body in a fixed deathly embrace; he could neither scream nor utter a cry; his chest muscles were locked tight too. Only tears moving slowly down his face told the story of his agony of mind as well as of body. He died on July 22, after a seizure so ferocious he was thrown halfway across the room.

The news of his death spread quickly. The *Brooklyn Eagle* lamented the terrible loss, and flags were set at half-mast. At a ceremony unveiling a statue of Roebling some time later, the speaker caught something of the man's indomitable spirit when he said: "I am expected to translate into words what the sculptor has so admirably expressed in bronze. But words are too plastic for such a task. As if his nature had been subdued to what it worked in, the Iron Master of Trenton was a man of iron. Iron was in his blood and sometimes entered his very soul." Was it possible now to build the bridge with the "man of iron" gone? He understood every aspect of the task. His experience, like gravity, held the dream in place and made it solid.

There was, however, some hope. His son, Washington, was also now trained as an engineer. After the Civil War, he had frequently assisted his father, learning and absorbing many of the older man's ideas and methods. They had been working together on the Cincinnati Bridge when the Brooklyn contract was won. In the last days of his life, while in terrible pain, John Roebling was reassured his son would endeavor to complete the huge task. An exchange of trust, a steady intention, and Washington Roebling was set on course. He would carry on his father's great work. In August 1869, he was appointed chief engineer by the board.

WASHINGTON ROEBLING'S FIRST task was to build the massive towers that his father had designed. It was essential that they be built on firm foundations deep under the East River. John Roebling had advised his son to use a new technology, pneumatic caissons—not unlike gigantic diving bells—enabling men to work beneath the river in the dry. Little was known about using the caisson, but there were rumors associated with them of a mysterious disease—"caisson disease"—which was very painful, sometimes fatal, and which haunted the caisson at deeper levels.

A caisson is like a large box, open at the bottom, that is gradually lowered onto the riverbed. Inside, it is pumped full of compressed air, stopping the river water from flooding in, thus enabling men to excavate the ground. As it is slowly lowered to reach solid rock, where it rests, the air pressure is increased to balance the extra pressure from outside. Roebling had visited Europe, where this technology had been developed, but this was the first time it was to be tried on such a scale and at such a depth. It was, said Roebling, "by far the largest structure of its kind ever sunk."

Over the next few months, the Brooklyn caisson took shape by the shore of the East River at the Greenpoint shipyard. It was a very large chamber, made of wood encased in metal, measuring 168 by 102 feet. The walls of the caisson were designed to facilitate cutting into the ground, nine feet wide at the top, tapering to a mere eight inches at the bottom. The lower walls were covered in iron, giving the sides a cutting edge, known as a "shoe." In March 1870, thousands came to view the launch of this strange vessel into the East River. A few weeks later, six tugs towed the caisson into its final position—the site of the old Fulton Ferry slip, where John Roebling had received his fateful injury. Once in place, the caisson's roof was strengthened with additional timbers, to 15 feet thick. Airlocks and man shafts were completed to enable workers to enter the chamber; other shafts were built to remove debris. Just above the waterline on the roof, the first courses of masonry for the tower were laid; this weight would help to lower the caisson. By early summer, excavation was ready to begin.

The workers entered the hot, humid underground world of the caisson through airlocks. Under the glare of calcium lights and at temperatures of 80 degrees and higher, a hundred men worked eight-hour shifts. Entombed below the river, they labored up to their knees in mud, slime oozing through the planks on the floor, the walls constantly saturated.

According to master mechanic Edmond Farrington, who had worked with the Roeblings on previous projects, it was like hell itself. "Inside the caisson, everything wore an unreal weird appearance," he wrote. "There was a confused sensation in the head, 'like the rush of many waters.' The pulse was at first accelerated and then sometimes fell below the normal rate. The voice sounded faint and unnatural and it became a great effort to speak. What with the flaming lights, the deep shadows, the confusing noise of hammers, drills and chains, the half naked forms flitting about . . . one might get a sense of Dante's inferno."

As they broke the rock and removed the sludgy, stinking material from the bed of the river, the sound of the workers' hammer blows became distorted and the smell of the waste hung heavily in the air. The mud and debris were placed below a water shaft where a scoop hauled it up to a waiting barge. When a few inches of earth were removed from the shoe, clamps underneath the partitions were knocked out, enabling the caisson to gradually settle lower. Progress was painfully slow at first, fewer than six inches a week. It was not possible to drive the caisson down just by the weight of the building work above, yet the boulders under the cutting edge of the caisson walls were laborious to remove.

The cutting edge of the caisson was not well sealed, often resulting in "blowouts" as the compressed air escaped. These sudden bursts of escaping air would send a fury of water and debris in a column as much as 500 feet above the river, only to descend in pandemonium on the workers who were placing the masonry for the tower. After every blowout, the bottom of the caisson was flooded, and a thick fog, hardly penetrated by the weak lighting, distorted both sight and sounds. No matter how absorbed the men were in their work, every minute of the day spent in that gaseous haze they were aware of the weight of water and masonry overhead and the fragility of their protection from it. "The noise made by splitting blocks and posts was rather ominous," wrote Roebling, "and inclined to make the reflecting mind nervous in view of the impending mass of thirty thousand tons overhead."

Inch by slow, painful inch, as the caisson was lowered, the large rocks, sometimes over 10 feet in length, had to be split and broken before the clamshell scoop in the water shaft could clear them. This would sometimes take days. There was one possible solution: use explosives to break the rock. This would buy time, but the consequences were unknown. It was entirely possible that these explosives could blow out one of the water shafts, which maintained the delicate balance of pressure in the caisson. In this situation, the compressed air would escape, the caisson would drop like a dead weight, and those within it could be crushed to death. The deeper the men went, however, the larger and more numerous the boulders became, until Roebling felt compelled at last to resort to the use of explosives. He experimented with small charges at first, horribly aware of the risks. Miraculously, boulders in the caisson were broken with no other internal change. As a result, downward progress increased up to 18 inches a week.

With so many naked flames, gaslights, candles, and explosives in use, fire was also always a hazard in the compressed air of the caisson. At 25 pounds per square inch, in this air the flame of a candle would reignite when extinguished. As they went deeper, guarding against fire was foremost in everyone's mind. Additional hoses and fire extinguishers were put in place and the lights were carefully monitored on each shift. Yet during the night of December 1, 1870, when the caisson was 42 feet deep, close to its final position, one fire escaped notice, with terrible consequences.

A worker holding a candle too near a small area of unprotected wood was thought to have accidentally started the fire, which ate its way hungrily into the solid wood of the roof and went undetected for several hours. The men panicked when the fire was discovered. Trapped below the river, the thought uppermost in every mind was to escape before the colossal weight of all the masonry above them came crashing down. In the pandemonium, the foreman, Charles Young, somehow established order and Roebling was summoned.

Because the fire was out of sight behind the roof timber, no one could be sure how long it had been burning and how much of the 15-foot roof timber had been destroyed. Edmond Farrington drilled holes into the roof to try to ascertain the damage. But each hole drilled fanned the flames of the tauntingly invisible fire with compressed air. If the fire was not put out, it would be just a matter of time before the enormous weight of the masonry crushed the men below. Flooding the caisson seemed the only answer, but this was as dangerous as the fire. If compressed air escaped before flooding was completed, a huge blowout would occur, the caisson would drop dramatically, and the damage would be irreparable.

Buckets of water were thrown at the roof and hoses were set up. By 5:00 in the morning the thick beams of the roof were saturated with water and the fire appeared to be out. Roebling, with unsparing dedication, had spent almost a day and a night in the caisson fighting to save the bridge. Now as the crisis appeared to be over, he suddenly collapsed. He began to complain of a certain amount of paralysis; he could barely move. Colleagues had to help him into the airlock, and then he was taken home by carriage. After only three hours' rest, however, the dreaded news came that the fire had not been defeated, that it was in fact out of control, high up in the roof. Exhausted, but unable to accept defeat, Roebling returned to the chamber. It was clear there was no alternative: they had to flood the caisson.

Everyone left as the vast caverns of the caisson were filled. It took seven hours. All the while Roebling stood by and watched. The unanswerable question was, would the fire eat through the timber before the water quenched the fire? If the fire won, that would be the end of the bridge. The press had a field day apportioning blame. The caisson was so severely damaged, declared one critic, "that I don't believe that any man living will now cross that bridge!"

Fortunately, all the skeptics were proved wrong. After two and a half days, the water was removed, the chambers were once again filled with compressed air, and the men returned to work in the dark, soaked world, 40 feet below the friendly daylight. The roof had held and, within a few weeks, the caisson reached bedrock. The vast chamber where the men had worked was slowly filled with concrete to strengthen the foundation for the tower. Success, at last, seemed assured.

All was proceeding well for 10 days, until, without warning, one of the supply shafts into the chamber jammed, causing a blowout. Material was flung out of the caisson like the eruption of Vesuvius, pouring debris out onto the men laying masonry above the waterline. They ran terrified from the scene, leaving the men below to what looked like certain death. Roebling himself was among the men trapped below.

"The noise was so deafening that no voice could be heard," he later wrote. "The release of water vapor from the rarefying air produced a dark, impenetrable cloud of mist and extinguished all lights. No man knew where he was going; all ran against pillars or posts or fell over each other in the darkness. The water rose to our knees and we supposed that the river had broken in . . . I was in a remote part of the caisson at the time; half a minute elapsed before I realised what was occurring and groped my way to the supply shaft where the air was blowing out." Several tense minutes elapsed before Roebling and his men succeeded in closing the door. The pressure inside the caisson had plummeted from 17 pounds to 4. Solid brick piers built to strengthen the roof had played a part in supporting the weight when, during the accident, 30,000 tons of pressure was unexpectedly applied. Remarkably, no long-term damage was done.

With the end in sight, to strengthen the roof before concrete filled the caisson, Roebling decided to fill the honeycomb made by the fire with cement. To his alarm, he soon found that although the cement was filling every hollowed track left by the fire, it was resting on charcoal, 1 to 3 inches thick in places, which stretched for up to 50 feet. This was a devastating discovery. It seemed the third and fourth courses of wood were pure charcoal! The enormous weight of the tower could never be allowed to rest on such fragile material. Painstakingly, all the cement had to be removed and every piece of charcoal scraped away before the hollowed roof was recemented. Finishing the caisson was delayed by three months and the cost was $50,000, but Roebling finally felt sure that the roof would be a firm support for the tower.

By March 1871, the foundation for the Brooklyn tower was completed. Now the vast structure could climb skyward for all to see. Roebling's annual report was eagerly awaited by the Bridge Company; he and his men were suddenly heroes to the public. Their efforts in the watery subterranean world, where unknown terrors were daily fare, took on the glory of fable in the press, especially since not a man had been lost or injured. Better still, there had been no real incidence of the mysterious caisson disease.

THE NEXT STAGE was to tackle the foundation on the New York side. Here the water was deeper; bedrock was thought to be around 100 feet below water level. This news was met with some apprehension, as caisson disease was always associated with greater depths. But much had been learned from the experience of the Brooklyn foundation and Roebling modified the caisson for the New York side. Because the chamber would have a considerably greater load resting on it, the roof was made 22 feet thick; and to protect against fire, the working chamber was lined with iron, and a large supply of water was made always available. The airlocks were also modified, making it possible for more than a hundred men to enter or leave the chamber at a time.

During September 1871, the New York caisson was maneuvered into position. A few weeks later, a large team of men were steadily excavating waste, under the supervision of engineers Colonel William Paine and Francis Collingwood. The caisson was resting on a site that had been a city dumping ground for decades, and the stench of rotting animal and vegetable waste was overwhelming; several men were overcome by the fumes. Dredge buckets could dispose of debris at a rate of 350 cubic yards of material a day, and lighter waste, such as sand and stones, was ejected by compressed air through pipes in the roof. More than 60 men worked hard to keep three pipes removing the sand and small stones. The deeper the caisson went, the greater the force at which the sand was ejected. Eventually, the pipes would blast material with great force 400 feet into the air. This was managed by turning the pipes in the roof at an angle to direct the fearsome column of sand and stone into waiting barges.

The downward journey of the unique vessel through mud, sand, and gravel was, at first, relatively easy as its weight alone helped to cut through the soft ground. At deeper levels, they encountered harder rock combined with quicksand where the dredge buckets could not be used, and the descent slowed. When the caisson reached a depth of 50 feet, the pressure in the chamber was 24 pounds and the working day was shortened. The men began to succumb to painful symptoms including headaches and severe leg cramps, which they described as like having flesh torn from every bone. By 60 feet, the pressure was 30 pounds. The number of men suffering from horrific symptoms was growing steadily. Paralysis, swollen joints, vomiting, double vision, and dizziness were common. "The labor below is always attended with a certain amount of risk to life and health," reported Roebling with some understatement. "And those who face it daily are therefore deserving of more than ordinary credit."

The Bridge Company hired the services of a Dr. Andrew Smith. He examined every worker to check on their physical condition, and kept careful records. For the majority of men, the attacks of caisson disease began when they left the pressurized chamber. It was soon nicknamed the "bends" as the onset would often cause a man to bend double with painful stomach cramps and vomiting. This was frequently accompanied by a fearsome pain in the legs, which could move rapidly up the body and end with a piercing headache, as if the patient had been "struck with a bullet." The men could feel dizzy and keel over without warning after being violently sick. Joints at the knees and elbows would be abnormally swollen. The face would be pale, like a death mask, and covered in beads of perspiration. Many suffered paralysis.

By the time the caisson reached a depth of 70 feet, the pressure in the chamber was up to 35 pounds; shifts were reduced to four hours a day. Morale was low, but bedrock was thought to be just another 30 feet lower. Dr. Smith took careful case notes, but the disease continued to baffle.

"Case 27, Patrick McKay," Smith recorded. "On the 30th April 1872 he remained in the caisson half an hour beyond the usual time. . . . When leaving, he was found collapsed with his back against the wall of the air lock, quite insensible." He was removed to a nearby hospital, where he remained unconscious. "Paroxysms of convulsions set in, in one of which he died, nine hours after the attack." Equally distressing was case 30, a German immigrant named John Myers. Myers had worked a two-and-a-half-hour shift, but feeling unwell had gone home. "As he passed though the lower story of the house on his way to his room, which was on the second floor, he complained of pain in the abdomen," recorded Dr. Smith. "While climbing the stairs, and when nearly at the top, he sank down insensible, and was dead before he could be laid upon his bed."

It was cases like these, where strong men were felled on a daily basis by an invisible enemy, that filled the men with a superstitious dread. Who would be next? A 16-year-old Irish laborer, named Frank Harris, described his first day: "In the bare shed where we got ready, the men told me no one could do the work for long without getting 'the bends,' which was a sort of compulsive fit that twisted one's body like a knot and often made you an invalid for life. We went into the air lock and they turned on one air cock after another of compressed air. The men put their hands to their ears . . . the drums of the ears are often driven in and burst if the compressed air is brought in too quickly. . . . We were working naked to the waist in temperatures of 80 degrees Fahrenheit and all the time we were standing in icy water that only was kept from rising by the terrific pressure. No wonder the headaches were blinding." Harris had intended to stay at least a month, but after seeing a man fall to the ground in agony with blood pouring from his mouth, he left immediately.

Mystified as to how to treat the disease, Dr. Smith advised the men to eat well, with plenty of meat and no alcohol. His range of remedies was restricted to whiskey, ergot, a mild painkiller, an antispasmodic injection and, as a last resort, morphine. Dr. Smith noted that symptoms were often experienced on leaving the pressurized area, and he speculated correctly that a longer time spent passing through the lock might be beneficial. Indeed, he even considered "that if sufficient time were allowed . . . the disease would never occur."

It is now known that caisson disease, or the bends, occurs after decompression, when nitrogen, which is dissolved in the blood, becomes gaseous, creating bubbles that can harm the tissues of the body by blocking the oxygen supply. Nitrogen bubbles can form anywhere—hence, the diverse symptoms—but if trapped in the nerves or spinal cord, they produce signs of paralysis. Dr. Smith was indeed correct to recommend slower decompression, which would allow the nitrogen gas to form slowly enough to be removed by the lungs. However, no one understood the significance of his conclusion at the time so his advice was ignored.

Roebling was suffering from painful symptoms, too. Working 14 hours a day, 6 days a week, he had put himself under great strain for many months. His wife, Emily, became increasingly worried at his physical decline as he struggled with the endless burden of leading the men in such dangerous conditions. Conscientious, responsible, solving problems before they could develop, holding in his mind every detail of the bridge and its construction, Roebling came close to exhaustion. The need to find bedrock was desperate.

At 78 feet, as the caisson inched lower, soundings were taken to discover the nature of the rock below. The picture that emerged was of an immensely hard, irregular area of rock covered with equally hard compacted sand and gravel. Roebling was facing a decision. Could this strata possibly be the resting place for the caisson? Geologically, this rock had never been disturbed by man. The surface was so hard it was almost impenetrable, breaking every tool used to attack it. He knew that if he carried on, insisting on finding bedrock, it could take up to a year. How many more men might be sacrificed?

Fresh cases of the bends were occurring daily. This very real danger added a ghoulishly frightening element to this already frightening workplace, where every man knew that a small accident or miscalculation could lead to a terrifying death, buried alive deep under the river. During May 1872, several new cases of caisson disease were reported, and another death. "Reardon, England, 38, corpulent," noted Dr. Smith. "Began work on May 17. . . . Immediately upon coming up from the second watch, Reardon was taken with very severe pain in the stomach, followed by vomiting. . . . The vomiting continued all night and towards morning he was taken to Centre Street Hospital where he gradually sank and on the 18th died."

Roebling made his decision. Digging was stopped. The caisson for the New York tower had found its resting place; it was not bedrock, but it was near enough. Roebling was not a man to gamble, but if he was wrong, the tower would be unstable; it could move sideways like the tower of Pisa and pull the entire bridge with it. The stakes were high, but he was convinced he had taken the right course and gave instructions for the caisson to be filled with concrete.

While the layers of concrete were being put in place in the early summer of 1872, Roebling himself was taken ill again with the bends. This time he completely collapsed and had to be taken back to Brooklyn, where it was thought he would not last the night. Emily sat with him as he lay close to death in the very house in which his father had died. "Relief from the excruciating pain was given by an injection of morphine directly in the arm where the pain was most intense," he said. Ultimately, his only release was complete "stupefaction by morphine" taken internally until the pains abated.

Just three years after the bridge had claimed his father's life, it seemed that it would now claim the son, who was just 35. And the bridge was not even half completed. Although the foundations below the river were in place, aboveground the main visible sign of success was the Brooklyn tower, now climbing almost 100 feet above the river. There were still years of work ahead before the bridge could be transformed from a dream in one man's mind to a reality in stone. And with the chief engineer so disabled, the entire project seemed in doubt.

FOR SEVERAL DAYS Roebling lay at home barely able to move or speak. But something indomitable in his spirit would not allow him to succumb to caisson disease or even to acknowledge it. Those around him saw him make an amazing recovery, almost as if by an act of will he refused to be ill. Just like his father, he possessed a tenacity that would not allow him to accept any kind of personal defeat. Although the disease dogged him throughout the summer and autumn of 1872, this was kept secret from the papers, and he managed to return to work to oversee the completion of the New York caisson and the building of the towers. Emily alone knew how much he was suffering and the lengths he went to conceal this from his men.

As the towers rose on each riverbank they dwarfed everything, as though they were part of Gulliver's world, and the urban sprawl around them merely dollhouses. The men, too, working the derricks and handling the blocks looked tiny, out of scale. Even the individual blocks for the tower were bigger than the men. The granite came from Maine, brought down by boat, a vast undertaking since an estimated 90,000 tons of granite and limestone were needed for the New York tower alone. At first, boom derricks handled the huge blocks, then when the towers became too tall, they were hauled up to the top by pulleys.

Although Roebling insisted on a daily inspection of derricks and working machinery so that any faulty equipment was spotted early, accidents inevitably occurred. Objects falling from a height were a constant worry; blocks of granite or small tools were all lethal weapons. There were several serious accidents and many more near-misses. Even if the cry of warning from above could be heard over the building noise or the constant wind, there was little time to get out of the way of falling debris. It did not do to be careless when guiding hoisting ropes onto engine drums, either: a hand could be caught and an arm pulled from its socket before help arrived.

In spite of the inherent danger, the men working up high on the towers never lacked confidence. They moved around freely, apparently with no concern that a false step could send them hundreds of feet below to certain death. The majority of accidents were caused by falling from a height—missing a step, misjudging wind pressure, or just plain carelessness. One poor man wheeled his barrow to eternity, clutching the handles as it moved off the plank into a 100-foot drop. On another occasion, a granite block was being hauled up by derrick when a guy wire holding two other derricks broke. Three men fell to their deaths, causing mayhem on the ground and crushing one man's skull like a matchbox. Another man, hearing the noise above, tried to jump clear but was crushed, sardined into a tiny space between masonry.

By December 1872, the Brooklyn tower stood over 140 feet high. The New York tower was lower at that point, but the promise of what was to come was tangible and public interest in the project was growing. With the onset of winter, however, Roebling finally had to acknowledge that he was desperately ill; even so, he could not bring himself to give up. Emily went to Henry Murphy, president of the Bridge Company, and explained that her husband needed a rest, but that he would never abandon his position as chief engineer. Persuasive and firm and with a quiet charm, Emily won Murphy's assurance that her husband could continue as long as all went according to plan. Yet even with the longed-for rest, Roebling's illness worsened; and as Christmas approached, the day came when, in spite of his fierce determination to continue, he was in such pain he could hardly get out of bed. There was no cure and little hope offered by doctors. Emily was told that he would never recover and that death was just a matter of time.

As the winter progressed, Roebling became increasingly depressed at the helplessness of his condition when there was so much to be done. Feeling that death might come at any day, and knowing that he alone held all the information and skills needed to see the bridge through to completion, he worked relentlessly from his sickbed. Ignoring his pain, he produced detailed designs for further work on the towers and the anchorages. His eyesight too was failing, and he wrote with great difficulty. When he became too weak to write, he dictated his directions to Emily, who, with his nerves ultrasensitive to the slightest sound, was the only person he could bear to have near him.

Winter slowly turned to spring and Roebling was still alive, still involved with the exacting minutiae of instructions for the bridge: information for stringing the cables, diagrams for the suspension cables and the radiating stays. But his doctors insisted on complete rest as his best hope of recovery. He had pushed himself tremendously that winter to complete all the plans for the bridge, and now, at last following medical advice, he and Emily went on a trip back to Germany, to the spa at Weisbaden, Saxony.

Roebling and Emily intended to leave the bridge for a couple of months at most, but they stayed abroad for six months in the desperate hope that complete rest would bring a cure. When they returned in the winter of 1873, he was still too much of an invalid to work again and, on doctor's advice, he went back to the family home in Trenton, New Jersey. It was from here, 60 miles from Brooklyn, that he continued to oversee the work on the bridge, sending out meticulous instructions to his colleagues on every detail of construction by letter or through Emily. He still had the same team of engineers, whom he knew and trusted and whose loyalty was beyond question. "Probably no great work was ever conducted by a man who had to work under so many disadvantages," Emily wrote. "It could never have been accomplished but for the unselfish devotion of his assistant engineers."

The towers continued to rise until they dominated the New York skyline. For those working at such heights it was a strange new experience. "There are times when standing alone one can feel completely isolated as in a dungeon," wrote Farrington. On one occasion, thick fog had risen to within 20 feet of the top of the tower. "I shall never forget that morning," he recalled. "The fog hung, dense, opaque, tangible, like a dull ocean obscuring all below." The spires of Trinity Church in New York and the tops of the masts of ships in dry docks in Brooklyn "were all that were visible of the world underneath. . . . Rising through this misty veil was the confused crash and roar of city life below."

During 1873, work began on the Brooklyn anchorage. The anchorages on either side of the river were to hold the four great cables in tension and were also part of the road approach to the bridge. Each massive anchorage, built in granite, was to be constructed 900 feet inland from the towers, making the total length of the bridge over a mile. The anchorages would house the four 23-ton anchor plates, which would be placed at the base of the anchorages in the early stages of building work, and then covered in masonry. The anchor plates were to hold the four cables running from the top of the tower tightly in place. From the center of each plate, 18 eye-bars 12 feet long acted as the first link in a long double chain of eye-bars that made its way in a curve to the top and the front of the anchorage. As each link was fastened to the next, it was buried under tons of granite while the walls of the anchorage rose to keep pace. By the time the mighty chain reached the top of the anchorage, there were 38 bars that would hold the cable wires in place.

The Brooklyn tower was finally completed in 1875, and attention turned to the New York side. Old tenements and warehouses covering the anchorage site were pulled down, to reveal a swamp and the original shoreline of Manhattan. Three steam pumps disposed of thousands of gallons of water before clear sand and gravel were revealed. The foundations were then driven into place. The granite walls progressed at a fast pace with the men working three shifts across 24 hours. By the summer of 1876, both anchorages were completed, virtually solid masonry rising 90 feet high and 119 by 129 feet at the base. They were massive mountains of granite, burying the anchor plates, which would grasp the cables and never let go.

Work for the bridge was now proceeding according to plan under the long-distance direction of Roebling, who was still living quietly in Trenton, seeing no one for days but his wife. He had the uncanny ability to anticipate problems that might arise on the bridge, and in spite of his fragile health, pushed himself to complete drawings, plans, and instructions. Although he never saw the work, he knew exactly what was happening. He held the whole concept in his mind. With his eyes closed he could envisage the complete anatomy of the bridge and flesh out the structure with every busy detail. He had no need to actually be there.

The household in Trenton revolved around the difficulties of Roebling's illness. He could not talk for long; his nerves would be excruciatingly tried by noise or even an unknown voice. The only voice he could bear to hear was that of his wife. "You know, darling, that your presence always made me feel so good," he had told her years before. "A kind of contented feeling pervaded me if you were only near." She became the link between him and the outside world. He relied on her absolutely and she never let him down. At a time when women were excluded from the engineering profession her knowledge and understanding of engineering was growing as she explained his instructions to the men on site. But even she could only feel daunted at the prospect of spinning the cables a third of a mile across the great East River.

BY THE SUMMER of 1876, both towers and anchorages stood ready to receive the cable. The next step was to make the first tenuous connection between the two towers, waiting like silent giants to catch the massive thread and spin the causeway. The process began by setting up "traveler ropes," and on August 14, 1876, a large crowd gathered to watch the first wire being stretched across the river.

Wire about an inch thick was placed on a reel in a boat stationed in front of the Brooklyn tower. The free end was taken up over the tower and secured to the anchorage. When this was done, the boat then proceeded to the New York tower, playing out the rope, which sank to the river bottom. On arrival it was taken to the top of the New York tower and then down again on the other side and attached to a hoisting engine at the bottom of the tower. The men had to wait until there was a break in the river traffic before the wire could be pulled tight. A cannon was fired to serve as a warning to ships, and the crowd began to cheer. In a matter of minutes the wire broke the surface of the water and fell into position, at around 200 feet above the river. The two towers, at last, were linked. Colonel Paine sent a telegram to Roebling to tell him the exciting news. Later that day, a second rope was secured and the two ends were joined to make a continuous traveler rope, which could be driven by an engine on the ground.

The time had come for someone to cross the East River on this first single wire 300 feet above the river, to prove that it was safe. Edmond Farrington, who had worked as an engineer with the Roeblings at Niagara and Cincinnati and had been the first to cross those great divides, did not hesitate. Roebling had great faith in his master mechanic: "He has the necessary coolness . . . and does not get easily frightened in times of danger." Farrington, strong, wiry, all of 60 years old, in his very best white Sunday suit and hat, set out over the river, sitting on a piece of wood attached to wire no more than an inch thick. It took some courage to trust such a narrow expanse so high above the river, but halfway across, undaunted, Farrington clapped his hands and waved his hat at the crowds below, who, delirious with excitement, had turned out in the thousands to watch him. Tugs and boats sounded their horns, church bells rang out, even a cannon was fired. "The ride gave me a magnificent view," he said calmly, "and such pleasing sensations as probably I shall never experience again."

During the autumn of 1876, Roebling was determined to move to New York to be nearer the bridge. He went to stay on West 50th Street with Emily's brother, General G. K. Warren. He was still far too frail to visit the bridge site, but now it was easier for the men to see him. Within a few months another traveler rope, a carrier rope, temporary wire cables, and a footbridge, four feet wide, were all in place. The footbridge was to be used by the workmen as they placed the cables. The men working up high against the sky making those first slender connections, often looking precariously balanced, just a misstep away from certain death as they spun the steel lace, always had a fascinated audience below.

The public, too, had a taste for excitement, and many thousands applied for tickets, issued by Henry Murphy, to cross the footbridge. Murphy himself decided against making the crossing since he felt that "the company could not yet afford to lose its President." As many as 70 a day were making the perilous trip on the constantly moving walkway.

Some were so terrified as to be a nuisance; others in their overconfidence were equally distracting for the workers. One person even wanted to make the trip on horseback. The final straw came when someone walking across had an epileptic fit. Several workers rushed to the rescue, holding the man down as he threw himself about on the narrow walkway 300 feet above the river. In desperation, they had to tie him to planks until the fit had passed.

Roebling, following his father's initial recommendation, was interested in using steel for the suspension wires, at that time entirely new and untried. Iron had always been used in the past, and no one was sure what strength of steel wire should be used. If the bridge failed, this would be an expensive experiment. Undeterred, Roebling arranged for samples of steel wire from different manufacturers to be tested. He found that the new material was far superior to the iron wire his father had used; he could obtain almost double the strength: 160,000 pounds per square inch. For greater durability and to prevent rusting, he also decided all the wire was to be galvanized.

Much discussion ensued among the board members about who should win the lucrative contract to make the cable. Abram Hewitt, vice president of the Bridge Company, insisted that bids from any company in which a trustee or engineer of the bridge had an interest could not be considered. Consequently, the family firm of John A. Roebling's Sons, making steel wire in Trenton, could not bid. Hence, Roebling decided to sell his stake in the family business so his brothers could try to win the contract. The Roeblings submitted the lowest quote: 6.75 cents per pound for steel wire. Despite this, the contract for 3,500 tons of steel wire was awarded to a Brooklyn firm run by J. Lloyd Haigh.

Roebling had his doubts about this firm, and about Haigh in particular, since it was known locally that Haigh had a somewhat questionable reputation. A resident of Columbia Heights for some years, the good-looking, charming, regular churchgoer had also been busy courting the ladies. He had been engaged to a local beauty until a Mrs. Haigh and her two children had appeared. Several more engagements followed with various young women, and Haigh even succeeded in marrying two in quick succession. His reputation as a businessman was rumored to be as interesting as his private life. Roebling, full of suspicion, made arrangements to test the wire from Haigh's yards, and even alerted his men to the possibility that Haigh might try to bribe the inspectors. He soon found out that Abram Hewitt had a mortgage on Haigh's Wire Company, which he would do nothing to redeem as long as Haigh kept paying him a significant percentage of his money from the wire contracts. Roebling's suspicions would be borne out later.

In the summer of 1877, five years after his first attack of caisson disease, Roebling and his wife returned to their house in Columbia Heights, Brooklyn. At last he could see the Great Bridge from the window of his house and watch through a telescope as spinning was started on the cables. Large drums holding 10 miles of wire each were placed on top of the Brooklyn anchorage. The end of the wire was looped around a groove in a large carrier wheel, which was hanging from the traveler rope that Farrington had crossed earlier. The wire was made fast to a strong iron brace, or "shoe," fixed in the Brooklyn anchorage. The traveler rope then took the wheel over the two towers to the New York anchorage, where the wire was attached to another shoe. As the wire around the carrier wheel formed a loop, two wires at a time, in fact, made the journey. This process was repeated endlessly, with men stationed high above the river en route, guiding each wire, so that all the wires hung with the same tension. It took 280 wires to make a strand, and 19 strands were used in a cable. These were then tightly bound and wrapped with more wire. When two strands were completed, the shoe was lowered and attached to the eye-bar on the anchorage, and a steel pin was passed through the eye-bar and the shoe.

For several months work proceeded well, until in December 1877 an incident brought growing criticism to the bridge trustees.

One of the arches that formed part of the approach road to the anchorage on the Brooklyn side of the bridge suddenly collapsed. Many tons of masonry fell and some workmen just missed being buried alive. In the confusion, the gathering crowd thought that all the arches were about to fall, maybe even the tower, too. Rumors spread rapidly that many men were buried under the rubble. It took some hours before it was established that only one man had died. At the inquest, it was found that the supporting structure for the arch had been removed before the cement had dried. It was a simple human error; nothing was fundamentally wrong with the bridge. The overworked engineer who was in charge of the Brooklyn approach, young George McNulty, was to blame. Yet the press would not be silenced. As chief engineer, Roebling was the man with the final responsibility for safety. Where *was* he?

No one, except the engineers in charge, had seen him for years. Was this the way to build a bridge, asked the critics? And such an extraordinary bridge as this? Rumors circulated that Roebling was blind, that he could not speak, that he was paralyzed, that his untrained wife was really in charge of the bridge. "It has become the deepest of mysteries . . . where the chief engineer is," declared the *Union* newspaper. "He may have been dead or buried for six months. He is surrounded by clouds impenetrable. . . . We declare the great East River Bridge in peril, because it has no head, because its wires of control run into somebody's closely guarded sick room."

On January 31, 1878, a severe gale destroyed ships and property in New York, but the bridge suffered no damage at all. The damage of public opinion was, however, having a far more devastating effect. Leading the criticism was the New York Council of Reform, whose members campaigned to stop work on the bridge entirely. They brought their grievances to court, arguing that the bridge was not only dangerous, but would damage commerce by obstructing navigation on the East River. The bridge had so far cost $13.5 million and it was a criminal waste to "squander" any more public money on this "highly experimental bridge," especially since, in most cases, "suspension bridges have been destroyed within a few years of their erection."

This anxiety was fueled by another accident on June 14, 1878. At lunchtime, workers on the New York tower were aware that something was wrong. To those on the ground the events were both frightening and confusing. Masonry was falling, a strange unidentifiable hissing noise split the air, and a strand was careering about in a furious frenzy. A man lay dying near the anchorage, another was dead, and two others were clearly in agony on top of the anchorage. Farrington himself was flung to the ground, but miraculously had escaped injury. There was panic in the streets as people fled.

The disaster occurred when a holding rope broke as a strand was being lowered into position on the New York anchorage. It swept away from the pulley and leaped hundreds of feet into the air, knocking over men who had been handling the pulley. It landed momentarily in a yard, only to be jerked up and sent in a mad course over the tower, before plunging down to the river, sending spray 50 feet up into the air and soaking passengers on the nearby ferries.

It was a difficult time for the Bridge Company. Political schemers in New York were busy spinning a web of intrigue. The authorities in New York refused to pay the money owed to the company. The controversy lasted throughout the early summer of 1878, with each day bringing more difficulties; eventually, funds ran out. Henry Murphy had to halt the work on the bridge, and men were dismissed, apart from a skeleton staff retained to work on the unfinished cables.

Then, in July 1878, one more problem was brought to Murphy's attention, and it appeared unsolvable. Roebling's men discovered that J.

Lloyd Haigh, who was supplying the steel for the cables, had been supplying inferior wire to the Bridge Company—possibly for several months. Inferior wire had been deliberately woven into the very fabric of the cables. Thus, it was impossible to remove and impossible to judge the safety of the cables. No one knew for sure what weakness had been woven into the Great Bridge.

COLONEL WILLIAM PAINE was in charge of inspecting the quality of the steel supplied by J. Lloyd Haigh, a man well-connected with senior board members. When a load of wire left the factory, he made arrangements for it to be tested, and if approved, given a certificate. Suspicious that substandard wire was secretly being sent to the bridge, Paine marked the wire that was approved. Sure enough, unmarked wire turned up at the bridge. Challenged about the fraud, Haigh assured everyone that it would not happen again. Cable making continued, and yet again the wire gave cause for alarm.

This time Paine and his men decided to follow Haigh's wagon en route to the bridge. They watched in disbelief as it was driven to the warehouse, unloaded at speed and refilled with "failed" wire, which was taken on to the bridge. The "certified" wire was then taken back to the factory to pass the test again. It was impossible to determine how long Haigh had been supplying inferior wire to the bridge—some estimates suggested as long as six months.

As Roebling had suspected, Haigh clearly could not be trusted. Henceforth, Roebling took steps to ensure that there was always a man on horseback escorting any load of wire coming from Haigh's factory. Haigh was then made to strengthen the cables, at his own expense, with 150 more wires that hadn't been specified in the original design. Roebling calculated that Haigh must have made $300,000 from the wire fraud; but in spite of problems arising from Haigh's greed, he reassured the bridge trustees that the cables were strong. In his original drawings for the cables, Roebling had designed them to be six times stronger than was necessary; now they were merely five times stronger.

Six months after Henry Murphy had laid off the men, the financial problems eased and work restarted. A bill was submitted and legislation passed that finally extracted the money that was owed by New York. Meanwhile, Roebling was preoccupied with the cables. Their strength depended on the exact positioning of each strand in the cable and this proved an exacting and time-consuming job. "If left but half an inch too long or too short for its true position," he said, "it will be too slack or too taut for its fellows, and it will be impossible to bind them solidly in one mass and make them pull equally together." It took over a year to spin the cables, but by the autumn of 1878 all four were finally complete. Each cable was 15 inches in diameter and 3,578 feet long, containing over 3,500 miles of wire.

As Roebling viewed the bridge through his telescope from Columbia Heights, he became increasingly convinced that for extra strength the suspension wires, which ran from the cable to the suspended roadway, should also be in steel. This had never been attempted before; the original plan had been to use wrought iron. The Brooklyn Bridge would be the first bridge made almost entirely of steel. Once again doubts were expressed in the papers, which were exacerbated by the news of another suspension bridge disaster. In December 1879, the magnificent bridge over the Firth of Tay in Scotland was demolished in a storm while carrying a train. Seventy-five people had died.

In spite of the continuing worries, the trustees did approve Roebling's plan to use steel for the suspension wires, which soon were seen hanging from the cables, each one capable of bearing a weight of 50 tons. There was always an audience at ground level; upturned faces absorbed in the delicate web being spun high above the river. As the suspension wires were fitted along the bridge with exquisitely exact spacing within the perfect curve of the cable, the bridge at last took shape.

With work on the suspension wires proceeding so well, attention turned to the floor of the great East River Bridge. Here, the floor beams were not made of solid steel, but of an open latticework design, which imparted greater strength. The major trusses were bolted to the suspender cables about seven feet apart and in line with the towers. The space between was filled in with more complex openwork steel. On top of this, six big trusses of steel ran from tower to tower, carrying the floor of the bridge. For a while, work had to be halted since the contractors could not keep up with demand for these innovatively designed trusses. But by December 1881, the floor beams were finished and a temporary plank path was installed so that the trustees and engineers could honor Mrs. Emily Roebling by accompanying her across the bridge; it was the first official crossing at road level.

Roebling himself was still too ill to leave his home. His wife had for so many years been his representative: his right arm, his good eye, his voice. Now all he could do on this momentous occasion was watch through the telescope as she crossed the bridge over the swirling water on a gray December day. When they reached the New York tower, glasses of champagne were raised to toast the success.

Meanwhile, Roebling had been asked by the trustees to strengthen the bridge so that it could carry the heavier load of training and freight. More headroom would be needed, more bracing required; this meant an extra 1,000 tons of steel. As the news leaked out that the bridge was to be strengthened further, more criticism appeared in the press. Certain political factions in New York wanted to know on whose authority the change had been made. The new young mayor of Brooklyn, Seth Low, argued that the delay and extra cost was unreasonable and it was now time to dismiss the absent chief engineer entirely. "I am convinced that at every possible point there is a weakness in the management of the Brooklyn Bridge," he insisted. "The engineering part of the structure—the most important—is in the hands of a sick man." Was the design of the bridge wrong? If not, why was money being wasted? If the bridge now had to carry all that extra steel, would it not collapse? Would it need the support of another pier? The criticism lasted for months, and rumors about Roebling's mental and physical abilities would not go away. Eventually, the trustees requested the presence of their chief engineer at the next meeting in the summer of 1882.

Roebling decided not to be "dragged to the board and put on exhibition," a sacrificial lamb for various ambitious political factions. He replied by letter: "I am not well enough to attend the meetings of the board as I can talk for only a few moments at a time, and cannot listen to conversation if it is continued very long. The physicians hope that living out of doors, away from the city, may lessen the irritation of the nerves of my face and head. I am now able to be out of my room occasionally. . . . I believe there is not a day when I do not do some sort of work for the Bridge. . . . My assistants do the work assigned them with perfect confidence . . . the interests of the Bridge do not suffer in any way from my not being there. . . . I shall be most highly honored to be present at meetings of the board when I am well enough."

Behind the scenes, there was a faction within the trustees intent on ousting Roebling, and when the trustees met in August it was decided that Cyril Charles Martin, the current second in command, should be given the post of chief engineer, with Roebling demoted to the role of first assistant. This was a major blow to Roebling. There was barely nine months' work left to do on the bridge. Was he to be sent packing ignominiously after 13 long years of conscientious endeavor on his father's extraordinary design? Roebling confided to a friend: "I have always had bitter enemies on the board for no reason except that I was in the way of any schemes for robbery."

He decided to write a long letter to the American Society of Civil Engineers outlining his work on the bridge and why he should not be dismissed. After many dictations to Emily, he arrived at a final statement, and she took it, in person, down Third Avenue to the institute to read it to the distinguished gathering. Usually women did not enter these hallowed halls and voice opinions on engineering; she was the first woman ever to address the society. The fact that she had come because her husband could not, coupled with her obvious passion for his cause and the rightness of his plea, all prompted a resounding response. "It produced an immense sensation," reported a correspondent for the *New York Times*. "It was a splendid statement and well delivered by Mrs. Roebling, who was young and handsome." The engineering profession had been totally won over.

However, Low was still determined to remove the chief engineer. He went to see Roebling in person to urge him to resign. Roebling refused, demanding to know the grounds for this request. Low could not really provide an argument but insisted that if Roebling would not go, then the board would fire him. According to Roebling's papers, Low's answers were so "weak and childish" he soon gave up any pretense at a rational argument and declared, "Mr. Roebling, I am going to remove you because it pleases me!" and turned abruptly to leave without another word.

At the next meeting of the board of trustees, votes were cast to decide on the chief engineer. The meeting was a long one, held behind closed doors, and the board was fairly evenly divided. Despite the political maneuverings of Low and his associates, there were several older trustees who were full of respect and admiration for Roebling. "As a bridge builder he has not had his equal on the face of the earth," declared one loyal supporter, William Marshall. Rallying Roebling's allies, he turned on Low and asked by what authority he demanded Roebling's resignation. "For one I would take the arm off my shoulder before I would permit myself to vote against a man standing here without blemish upon his character or ability." When the doors finally opened, the decision was made. By 10 votes to 7, Washington Roebling would remain as chief engineer.

In the early months of 1883, the bridge was nearing completion. Diagonal stays radiated from the top of the tower to the bridge floor, and vertical suspension wires gave further strength. The whole intricate mesh imparted a fascinating mathematical beauty. A wide promenade, 18 feet above the roadway, was in place. The approaches to the bridge were almost finished and plans were put in place for cable cars to start operating that autumn. Finally, an opening date was set for May 24, 1883.

The grand opening was to be a huge celebration. Special invitations were elegantly engraved by Tiffany's. Dignitaries and officials were invited from all over America, including President Chester A. Arthur. The crowds from out of town started at first light. It had become a party day for New York. Even though the bridge would not be open to the public until midnight, by midmorning the streets leading to the river were so tightly packed the crowds could hardly move. Those resplendent in their Sunday best jostled with country farmers, smart New Yorkers, city urchins and street vendors. The excitement was infectious. Flags fluttered everywhere: from the shops, the houses, and the boats thronging the river, and from the bridge itself.

At midday, the opening ceremonies began. Countless important officials, citizens, and members of the press promenaded across the bridge in the bright May weather. The three hours of speeches that followed were full of praise for the magnificence of the bridge. The Honorable William Kingsley, speaking for the board of trustees, declared "with one name this Bridge will always be associated—that of *Roebling*." He talked of how John Roebling had sacrificed his life for the bridge, and how his son too, "who entered our service young and full of life, hope and daring," had been struck down by the mysterious caisson disease and had never seen the bridge, save from a distance. "Roebling may never walk across this bridge as so many of his fellow men have done today, but while this structure stands all those who use it will be his debtors." Abram Hewitt went further: "During all these years of trial, and false report, a great soul lay in the shadow of death, praying only to stay long enough for the completion of the work to which he had devoted his life. I say 'great soul,' for in the spring time of his youth, with friends and fortune at his command, he braved death and sacrificed his health to the duties which had devolved upon him as the inheritor of his father's fame and the executor of his father's plans."

Roebling himself remained absent throughout the ceremony, silently viewing the proceedings from a distance. From his familiar vantage point at the window, he could see the toy-size figures of the dignitaries, made remote by distance, and hear the sounds of the band carried up on the breeze. The bridge was the most significant structure in the landscape: the combination of utility and beauty always surprising.

In Columbia Heights, Emily gave a small reception for a few friends and notables. The house was decked with flowers—lilacs, lilies, and roses in every room—their heady perfume everywhere, so that it seemed as if the garden had entered the house. She had placed a small marble bust of her husband in the parlor and crowned it with a laurel wreath and the words: "Let him who has won it, bear the palm." Washington and Emily greeted their guests together. He hardly spoke, but he shook hands with the president, who bore a large smile and a lot of praise for the bridge; then he retired quietly to his room.

By evening, the celebrations were far from over. There was to be a fireworks display of incredible extravagance. The crowds of earlier had now doubled to watch as the afternoon turned to velvet dusk and the sky was made dazzling by brilliant rockets, showers of colored stars, golden rain, and exploding bombs of color in abandoned profusion. The crowds were euphoric and shouted themselves hoarse as every band, trumpet, and instrument in the vicinity joined in. As the last fireworks died away, the full moon rose, bathing the bridge in mysterious light.

The bridge was now born; it had a life of its own, and began collecting praise like a magnet. New York—indeed, the whole of America—was thrilled with the Brooklyn Bridge, in love with the sheer American audacity of the enterprise, with the heroism cemented into its very fabric, in love with the man, who against the odds, had built it. It had become the eighth wonder of the world; people jumped from it, fell in love on it, married on it, someone even took a herd of elephants over it; and by 1888, 30 million passengers a year took the train over it.

In the 14 years it had taken to build, the Great Bridge had claimed the lives of 20 men. John Roebling, too, had exchanged his own life for his all-absorbing dream and, unknowingly, had condemned his son to a shadow life. Yet, as he had hoped, the bridge now stood a triumphant symbol to what men could achieve. Indeed, it had become such an icon it was seen not only as "an embodiment of the scientific knowledge of physical laws," but also as a "monument to the moral qualities of the human soul."

A small plaque at either tower commemorates the two men, father and son, who had struggled so fiercely for the bridge. Later, a plaque was added for Emily. Some 25 years after the inaugural event, when the bridge had defied outrageous storms and icy winters, all petty criticism forgotten, and even the struggle of the men who built it now part of history, a statue of John Roebling was unveiled. "Yes the bridge is beautiful," said the speaker. "His genius stands embodied in this form. . . . Whatever of faith, feeling, reverence John Roebling cherished in his heart, was here voiced like a ringing cry." And that ringing cry has echoed down the decades as the bridge stands, a testament to the human spirit.

CHAPTER 4

THE LONDON SEWERS

It is . . . the most extensive and wonderful work
of modern times . . .

OBSERVER, *April 1861*

DR. JOHN SNOW was a careful man, not given to extravagant claims or wild speculation. Cold, hard facts were what he valued above all, and his desk in his Soho surgery in central London was filled with the treasure of years of patient observation. Lists, letters, and diagrams lay everywhere, each covered with meticulous notes in the young doctor's spidery handwriting. Lines crisscrossed the pages, linking names to addresses; small maps of London streets were marked here and there; reports lay side by side for easy comparison. What the notes showed were connections—connections between individuals, linked in life by their neighboring homes and, finally, by a common cause of death: the cholera.

Cholera—the very word terrorized, conjuring up the prospect of a terrifying and sudden death, often within hours. The early signs, debilitating diarrhea and vomiting causing severe dehydration, were followed all too soon by kidney failure and collapse, stilled only by death. The disease emptied the houses and streets of those wealthy enough to run from its mysterious presence. Impossible to know where it would linger next; one row of houses untouched, the next all victims to its deadly embrace.

Cholera had come to fascinate John Snow and he aimed to understand its cause and means of transmission.

Since opening his Soho practice 10 years earlier in 1838, and before this, while serving his apprenticeship in Newcastle-upon-Tyne, the young doctor had followed cholera's unstoppable progress. The disease had arrived on British shores in the autumn of 1831 after a 15-year journey, which from its beginnings in India had traveled slowly along the trade routes like a fungus, leaving its mark in the cities of China and Russia, before finding its way at last to Europe and London. During the British epidemic of 1831, Snow had seen at first hand the devastation wreaked by the disease. Many times he had been a helpless witness to frightened patients in the final stages, with their puckered lips and blue sunken faces. It was a traumatic way to die, and by the time the 1831 epidemic had come to an end, more than 6,000 Londoners had succumbed.

The medical establishment around him was at an utter loss. "We know nothing," reported the medical journal the *Lancet,* "we are at sea in a whirlpool of conjecture." Amid the confusion, various theories sprang up. Some believed the disease was spread by atmospheric electricity, or that it rose from the earth, or even that "moral depravity" contributed to a person's vulnerability to disease. Without knowledge of cause or cure, officials were powerless to check the spread of the disease. And, now, as 1848 drew to a close, Snow found himself at the heart of a second, more serious cholera epidemic.

Although a leading anesthetist, Snow devoted himself to developing his own theory of cholera contagion, moving closer to a conclusion with every day that passed. He felt certain he alone knew why thousands were dying of cholera all around him. He alone knew how it stole upon its victim so innocuously, soon to be racing through the blood, delivering destruction. Even as he struggled to marshal the facts that would prove his theory beyond all doubt, the death toll rose. In London alone 30,000 people contracted the disease; there had been nothing like it since the bubonic plague nearly 200 years before. As panic spread, attention turned to the appalling unsanitary conditions of the capital and, in particular, the squalid living conditions of the poor.

In the first half of the nineteenth century, the population of London soared to 2.5 million, as thousands flocked to the capital to find work. Poor families often lived eight or nine to a single room of no more than seven feet square, in lodgings resembling stables more than homes. They cooked, washed, slept, and defecated within these four walls, and as the population grew, so did the smell. Foul odors emanated from more than 200,000 cesspools across London, in alleyways, yards, even the basements of houses. It was not a smell that could be easily washed away.

Traditionally, London's cesspools were cleared by the "nightsoil men" who carted sewage to outlying regions and farms. The nightsoil men worked in teams of four. After taking up floorboards or flagstones and placing large horn lanterns at the entrance to the cesspool, a "hole-man" would descend a few feet into the pit to fill his tub. He would then be helped by a "ropeman" to raise the stinking cargo, hopefully without too much spillage, and the waste would be emptied into carts by two burly "tubmen," carrying well over 100 pounds of sewage. As work progressed, the holeman would have to climb deeper with each trip, stirring up the filthy sludge to loosen it for shoveling out. It was an unbearably vile job for which the only compensation was a high wage, often up to two to three times the salary of a skilled man.

This system had worked well for many years, but as the city expanded, the journey to the outskirts of London became longer, and the nightsoil men's prices soared. At a shilling per cesspool, many Londoners could no longer afford the services, so raw sewage began to accumulate in the dwellings of the poor. The black suppurating cesspools hidden in the cellars were seldom emptied. Their contents were allowed to ooze through the floorboards or cracks in walls and flow at random through yards and ditches, fouling everything they touched.

Though there were sewers in London, these had originally been developed for an entirely different purpose: to serve as drains for surface water only, not for human waste. The sewers had evolved over the centuries using the natural drainage of streams and rivers that flowed into the Thames. Eventually, they had been covered so they would act as storm drains only. But as the population grew and the city's cesspools started to overflow, by sheer necessity, in 1815, the ban was lifted and sewage found its way through the old drains and sewers into the Thames.

The pressure on London's 200,000 cesspools and the haphazard sewer system was exacerbated by the new "must-have" accessory of late Georgian England: the water closet, the toilet. Developed commercially by a cabinetmaker, Joseph Bramah, who patented his design in 1778, the water closet's popularity spread rapidly throughout the capital. With each new flush lavatory adding 160 gallons of waste and water, London's sanitation soon reached crisis point.

There was so much sewage in London that an entire underclass, driven by grinding poverty, tried to scrape a meager living from it. The city's poorest, less fortunate even than the paid nightsoil men, struggled to get by as street sweepers, bone-grubbers, or "finders" of one kind or another. There were the "bunters," women who collected dog excrement from the streets for use in the tanning process, and the "toshers" or "shoremen," named for the "tosh," or scraps of copper, they recovered from the bowels of the sewers themselves.

Toshers could be seen stooping along the shore of the Thames in their "long greasy velveteen coats, furnished with pockets of vast capacity, their nether limbs encased in dirty canvas trowsers," observed investigative journalist Henry Mayhew in his social survey, *London Labour and the London Poor*. Typically, they carried with them the tools of their trade, a lantern strapped to their chest and a long pole with a large iron hoe at one end, which was used to test the ground before them so they didn't sink deep into a quagmire of sewage. Inside the old sewers the risks increased. Some toshers were said to have lost their way in the rancid passages until, overpowered by the appalling stench, they dropped down and died on the spot. Others were buried under sudden avalanches of old bricks and putrefying matter that could collapse into the passages. Rats, too, were a common threat and were particularly dangerous when cornered. Men working alone, it was said, could be "beset by myriads of enormous rats" and discovered days later "their skeletons . . . picked to the very bones."

For this reason, toshers often worked together, wading through the sludge in the dark underground tunnels, raking through the muck with their hoes for copper, iron, rope, or bone, all of which could be sold for a small sum. "There's some places, 'specially in the old sewers, where they say . . . the foul air 'ill cause instantaneous death," one tosher told Mayhew. "When I'm hard up," he continued, "well then I knows as how I must work, and then I goes at it like a stick a breaking." Despite the conditions, when "in luck" as they called it, the men would return home with a silver spoon, fork, or knife, or even perhaps pieces of jewelry.

At the very bottom of this chain, eeking out a living from human waste, were children. Invariably orphaned, or compelled by some other overwhelming calamity, they tried to appease their hunger from the proceeds of their finds in the mud and waste of the river. Known as "mud larks," since they could sometimes wade up to their waists into the mud and filth along the shore, they searched for the scraps even the toshers did not want. "These poor creatures are often the most deplorable in their appearance of any I have met with," reported Mayhew. "It cannot be said that they are clad in rags, for they are scarcely half covered by the tattered, indescribable things that serve them for clothing; their bodies are grimed . . . and their torn garments stiffened up like boards with dirt." They congregated in sewers close to shipyards, sifting for nails, bolts, lumps of coal or wood. Coal, when they found it, could be sold to the poorest in the neighborhood at a penny a pot, while iron, bones, rope, and copper nails were sold to rag shops for a pittance. "Should the tide come up without my having found something," one mud lark told Mayhew, "why I must starve till next low tide." Another woman explained that she had been too ill to work, so she depended on what her 9-year-old boy brought back from the river. "He never complained," she said, "and assured me that one day God would see us cared for."

As conditions grew worse, reformers became more outspoken over the urgent need for change. The lawyer Edwin Chadwick, secretary to the Poor Law Commission, was one of many to draw attention to London's unsanitary living conditions. In 1842, he produced an uncompromising and influential paper, "The Report on the Sanitary Condition of the Labouring Population of Great Britain." Shocked by the squalor of the slums, he cited "atmospheric impurities produced by decomposing animal and vegetable substances," "damp and filth," and "close and overcrowded dwellings" as leading inevitably to disease and epidemics. For Chadwick, the report marked the beginning of his own theory of cholera contagion—one that was to have terrible repercussions.

The theory increasingly favored by most leading practitioners and social reformers was the Miasma Theory. Disease, the miasmatists believed, was spread through the atmosphere like a poisonous gas; if the deadly smell that pervaded London like a mist was banished, cholera would be as well. Florence Nightingale was a well-known supporter of this theory, as was the influential William Farr who, backed by Chadwick, held the newly created office of chief statistician to the Office of the Registrar General. As such, it was Farr's job to register cholera deaths, and he had no doubt that "a disease mist, arising from the breath of two millions of people, from open sewers and cesspools, graves and slaughter houses," hovered "like an angel of death" over London. This disease mist, he wrote in colorful terms, could be pervaded in one season, by "Cholera, in another by Influenza; at one time it bears Smallpox, Measles, Scarlatina and Whooping Cough among your children." Edwin

Chadwick, too, was totally convinced by the miasmatists. His 1842 report pointed repeatedly to the defective ventilation and noxious odors in poor dwellings. "All smell is disease," he told Parliament confidently.

The key to these problems, Chadwick believed, was sanitary reform: to remove the smell from the dwelling place. In typically combative style he wrote in scathing terms of the system for collecting human waste, which, he declared, is a "vast monument of defective administration, of lavish expenditure and extremely defective execution." Convinced that smell led to disease, Chadwick and his colleagues were successful in introducing the new Cholera Bill, as it was known, which allowed the Privy Council to order existing dwellings to be connected to the sewage system, with big fines for noncompliance. This ensured that the already-choking byways and ditches now added their burden of excrement to the Thames and turned its once sparkling waters into one big open sewer, brown and sluggish with its unwanted cargo. Effectively, the Thames is made "one great cesspool," observed a leading developer, "instead of each person having one of his own."

Ironically, the very reform that Chadwick proudly believed was doing most to save peoples' lives was, in fact, putting them in great danger. While the poor drew their water from wells contaminated by overflowing cesspools, the middle classes enjoyed the convenience of drinking water piped directly to their homes, but this water was supplied by companies that had intakes on the tidal section of the river, which meant they were simply drawing sewage from 2.5 million people and rerouting it back into their homes. Although the water companies fiercely defended the quality of their water, comparing it with the "purest spring water," the implications of poor sanitation and misguided legislation were clear. Londoners were not only wallowing in their own excrement at home and in the streets, they were *drinking* a diluted concoction of raw sewage.

And sure enough, shortly after the new Cholera Bill was introduced, terror reigned once more on the streets of London. The second cholera epidemic of 1848 swept through London with even greater vigor than the first. In some poor districts, death rates were so high that the inhabitants could not afford even to bury their dead. "Nearly the whole of the labouring population have only one room," wrote Chadwick despairingly. "The corpse is therefore kept in that room where the inmates sleep and have their meals." So desperate were they to hide the smell from a rapidly decomposing corpse, which mingled with the usual foul background vapor, that they resorted to scenting the room with peeled onions. In the face of this second outbreak, the authorities were utterly at a loss.

Following an official enquiry, the Metropolitan Sewers Commission was created, in which the various sewer authorities across London were merged to form one single body charged with tackling the appalling sewage problem. Chadwick, as a leading figure in this First Commission, took a number of additional steps to try to control cholera by removing the smell. The overflowing cesspools were abolished, and the commission was given the power to check dwellings to ensure that the Cholera Bill was being implemented and that houses were connected to the sewers. In a further attempt to clean up the London streets and homes, he implemented a policy of flushing the sewers. Worried that waste could accumulate in the sewers beneath the streets and add to the putrid smell, flushers were employed to cleanse the sewers, dislodge coagulated deposits, and move the waste along and into the River Thames.

While the authorities concerned themselves with abolishing London's smell, one man was sure that, although the smell was an obvious clue to the cholera outbreaks, the deadly disease was actually lurking in, for example, a glass of drinking water. In 1848, the same year that the Metropolitan Sewers Commission was established, Dr. John Snow, who had been weaving together threads of evidence for some years, made a series of breakthroughs in his theory of cholera contagion. Water, he believed, played a leading role.

First, he was struck by a curious observation he made regarding the joint towns of Dumfries and Maxwell-Town. These were usually quite healthy places to live, he noted, but during both the 1832 and 1848 epidemics, many had died there. He discovered the inhabitants drank the water of the River Nith, a waterway into which he observed "sewers empty themselves, their contents floating afterwards to and fro with the tide." Could contaminated drinking water—rather than smell—be spreading cholera in some way?

Anecdotal evidence supported this idea. It was known that the sewage workers themselves, the flushermen, who spent their days immersed in the foul, airless sewers, had remained untouched by the current epidemic. Mayhew, who had interviewed a number of flushermen during the 1848 epidemic, was struck by this remarkable anomaly. "The cholera, I heard from several quarters, did not attack *any* of the flushermen [emphasis mine]," he reported. "The answer to an inquiry on the subject generally was, 'Not one that I know of.'" Yet if the disease was spread by smell, logically then, the sewer workers should be the first to die.

Snow, with his cautious, scientific approach was not, however, one to be swayed by anecdotal observations alone. Then, during the 1848 epidemic, he learned of intriguing new evidence. On Thomas Street, in Horsleydown, there were two blocks of poor housing situated back to back. On the north side, in Surrey Buildings, the cholera had caused many deaths, while on the south side, in Truscott's Court, there had been only one fatality. The only difference was that the inhabitants of Surrey Buildings poured their slops into a channel in front of the houses, which flowed into the well from which they drew their water. In fact, so much sewage found its way into the well that, when it was emptied, the entire base was encrusted with a black deposit. Snow realized that the people living on the affected side were drinking water contaminated with raw sewage.

This was an important piece in a complex puzzle, one that gave Snow enough material to draft a paper, in which he set out his own very different theory of contagion. Snow was sure now that the key was water. Cholera, he believed, was not spread by the stench of the sewers but "by the emptying of sewers into the drinking water of the community." In other words, it was not airborne but waterborne. And it had spread so successfully that by the end of 1849, cholera had claimed the lives of more than 14,000 in London alone.

Snow published the first version of his paper, "On the Mode of Communication of Cholera," at the end of the 1848–49 epidemic. So sure was he of his conclusions that he even went so far as to state that cholera could be kept at bay and "the enemy shorn of his chief terrors" by simple measures: People should avoid drinking any water "into which drains and sewers empty themselves." If that cannot be accomplished, the water should be "filtered and well boiled" to destroy the cholera agent. To avoid accidental contamination, anyone even attending a cholera patient "must wash their hands carefully and frequently," especially before touching food.

Snow's message sounded simple enough, but nobody was listening.

ONLY MINUTES FROM Snow's Soho surgery lay the office of another man whose life's work would center on the problem of London's waste. As Snow labored over his mounting paperwork in Frith Street, a young engineer by the name of Joseph Bazalgette faced similar problems at the offices of the newly formed Metropolitan Sewers Commission on neighboring Greek Street. Two remarkable men from very different worlds— Snow the son of a Yorkshire farmer who began his apprenticeship at 14;

Bazalgette privately educated and well connected: their lives would run in parallel but never intersect.

Like Snow, Bazalgette was not a man to spare himself, and his appetite for work had put a great strain on his delicate constitution. For several years during the 1840s, when railway mania was gripping the country, he had pushed himself as his offices on Great George Street in central London had been overwhelmed with demand. As he later admitted, he "nearly killed myself with hard work," leading a large team designing railways, canals, and countless other schemes. Newly married, his young wife, Maria, had watched helplessly as the pressure of work eventually caused his physical collapse and required him to spend a quiet year in the country to recuperate. Of slight build, it was Bazalgette's face that commanded attention, in particular his aquiline nose and keen gray eyes. All who met him were left with an impression of a man of exceptional power and quiet authority.

Bazalgette returned from the country in 1849 to find the city in the final throes of its second cholera epidemic. In the spring, he applied for a post as assistant surveyor to the Metropolitan Sewers Commission. Possibly hoping to impress his prospective employers, Bazalgette devised a detailed proposal to send in with his application. His suggestion was to provide public toilets throughout the capital to provide "relief from frequent personal inconvenience, and occasional pain, if not physical injury." With painstaking attention to detail that would later become his trademark, Bazalgette even spent time observing how people used existing urinals, then estimated potential profits. Each customer, he told the commission persuasively, would leave behind half a pint of urine and, by selling this on, each urinal could generate £48 worth a year, making an annual profit of £1,280, "a decidedly profitable speculation." But the Sewers Commission remained unmoved and his plan was not taken up.

At this time, there was growing criticism in the press, citing the London authorities' failure to tackle the sanitation crisis. The squalid living conditions of the city's poor became the focus of much debate; the sheer desperation of their plight had even reached the august readers of the *Times* when a famous letter, complete with 54 signatures, was printed in the widely read newspaper.

THE EDITUR OF THE TIMES PAPER

> *Sur,—May we beg and beseech your proteckshion and power.*
> *We are Sur, as it may be, livin in a Willderniss, so far as the rest of*
> *London knows anything of us, or as the rich and great people care*
> *about. We live in muck and filthe. We aint got no privies, no dust*
> *bins, no drains, no water splies, and no drain or suer in the hole*
> *place. The Suer Company in Greek Street, Soho Square, all great*
> *rich and powerful men, take no notice watsotnedever of our com-*
> *plaints. The Stenche of a Gully-hole is disgustin. We all of us suf-*
> *fur, and numbers are ill, and if the Colera comes Lord help us . . .*

Teusday, Juley 3, 1849

Well aware of the urgency of the crisis and anxious that no more time be wasted, Bazalgette reapplied to the Sewers Commission in August 1849. This time he was successful, and was duly appointed assistant surveyor to the second Metropolitan Sewers Commission. His first task was to familiarize himself with the situation. It was clear to Bazalgette that there was widespread disagreement among the different commissioners and that Edwin Chadwick had been a forceful influence. Numerous leading experts supported Chadwick's belief in the miasma theory and Bazalgette had no reason to question it. He, too, was familiar with Chadwick's report and recognized the truth of many of his observations: there was no coherent sewage system in place; work had been carried out randomly by common builders rather than qualified surveyors, and as a result, drains of different sizes and levels were not compatible. In an attempt to resolve these differences the Metropolitan Sewers Commission invited engineers and others to submit proposals to redesign the sewers.

On Bazalgette's desk sat a tidy stack of documents, representing a fraction of the 137 proposals sent in for consideration. There were countless different approaches. The news agent and M.P., W.H. Smith, proposed moving waste from the cities by rail; others suggested transport by barge; still others had schemes for deodorizing the sewage in situ. Bazalgette's first task was to assess these plans and report back to the commission as soon as possible.

By the spring of 1850, Bazalgette duly presented his findings to the recently formed third Sewers Commission. Twenty-one plans were thrown out as speculative and vague; the remaining 116 submissions were categorized with explanatory notes and presented for discussion by Bazalgette, with some misgivings. The committee, in some confusion over the range, intricacy, and frequent absurdity of the proposals, on March 8, 1850, threw out the lot as totally unworkable.

Over the next few years, one commission succeeded another while plans for the sewers remained snarled in bureaucracy. Chadwick was removed from the group, but even without his contentious influence, paralyzing disagreements between the different officials continued. In one long-running battle, a promising design by Frank Forster, engineer to the commission, was first rejected, then reconsidered, only to be rejected once again. While different factions argued over the scheme, in 1852, the defeated Forster died during the fourth Sewers Commission. "A more general support from the Board he served might have prolonged a valuable life," claimed his obituary, "which, as it was, became embittered and shortened, by the labours, thwartings and anxieties of a thankless office."

Next in line for this most stressful of jobs was Forster's deputy,

Joseph Bazalgette, who had been rapidly promoted. Bazalgette, quietly confident as ever, took up the challenge; but, as he set to work, an old enemy was waiting in the wings. In 1853, unexpected and unwelcome, the cholera returned to the city for a third time.

THE STREETS OF Soho, devoid of their usual lively ambience, were deathly silent, echoing to the slightest sound. Those who could had run from the unseen terror, taking their few possessions and leaving a ghost town of the dead and dying. Only a few remained, mostly men.

John Snow, too, remained, in Soho. On his desk lay a large map of the area. He had researched and prepared it with great care. The streets and houses were clearly demarcated, as were sewers, cesspools, and wells. In the center, a tight network of adjoining streets stood out amongst the rest. This small section was marked again and again with short black lines, the darkest cluster centering on Broad Street, near Golden Square. Each black line represented a death. Once again, cholera was stalking the streets.

But Snow was approaching a breakthrough. From the General Register Office, he had obtained a list of the deaths in and around Golden Square, Berwick Street, and St. Ann's, and had made house-to-house inquiries. He was extremely concerned that the mortality in this limited area "probably equals any that was ever caused in this country, even by the plague; and it was much more sudden." Within 250 yards of the spot where Cambridge Street joined Broad Street, there had been 500 fatal attacks in 10 days. Snow concentrated his attention on a particular localized outbreak in which 83 had died in the final three days of the week ending September 3. He found that all deaths had taken place within a short distance of a pump in Broad Street, not far from his surgery.

As the evidence accumulated, it became easier to decode the map and, in turn, the epidemiology of the disease. Ten deaths he noted were nearer to a different pump, so he decided to visit the relatives of the deceased in search of an explanation. They did not drink from their local pump, they told Snow, but had sent out for water from Broad Street, as they preferred the taste. In three further cases he discovered that the deceased were children who went to school near the Broad Street pump and were known to drink there. He also learned that water from the pump was sold in small shops with a teaspoon of effervescing powder called sherbet and often mixed with spirits in the local public houses, dining rooms, and coffee shops. Snow began to record the evidence. "The keeper of a coffee shop in the neighborhood which was frequented by mechanics and where the pump water was supplied at dinner time, informed me that she was already aware of nine of her customers who were dead."

He went on to investigate the anomalous survival rate at the nearby brewery. There was no mystery there; all 70 workers remained unscathed because they were provided with free beer or water from the brewery's own pump. He also followed up on seven workmen who lived outside Soho, all of whom had died. The unlucky crew, it transpired, had been working on Broad Street making dentists' materials, and regularly quenched their thirst at the Broad Street pump.

Most compelling was the case of an elderly widow who had died in Hampstead during the outbreak on September 2, despite not having been in the neighborhood of Broad Street for many months. However, Snow explained, a cart went from Broad Street to West End every day, "and it was the custom to take out a large bottle of the water from the pump in Broad Street as she preferred it. The water was taken on Thursday 31 August and she drank of it in the evening, and also on Friday. She was seized with the cholera on the evening of the latter day, and died on Saturday. . . . A niece, who was on a visit to this lady, also drank of the water; she returned to her residence, in a high and healthy part of Islington, was attacked with cholera, and died also. There was no cholera at the

time," he concluded triumphantly, "either at West End or in the neigh-
bourhood where the niece died."

Snow needed no further proof. On September 7, he took his evidence
to the Board of Guardians of St. James' Parish. The pump handle was
removed the following day and the outbreak was contained. Later it was
shown that a nearby sewer could indeed have contaminated the Broad
Street well. It was a victory, but a small one.

While this parish council was prepared to follow Snow's advice, few
others did. This was perhaps due more to Snow's character than his abilities. Though a genius in his field, he was extremely shy and ill at ease in
the company of others. In the little spare time he had, he preferred lone
pursuits, collecting geological specimens and walking in the country. He
never married and, for most of his life, was a teetotaler and a vegetarian,
who was regarded as odd by many of his contemporaries. As a result,
Snow never sought to become part of the medical establishment exempli-
fied by stronger characters like Edwin Chadwick and William Farr, and
consequently was not accepted by it. These were the men with the power
to take Snow's theory further, but both were far too entrenched in the
closed logic of miasmatism.

Nevertheless, Snow continued to gather his body of medical and
anecdotal evidence and put his case forward as often as possible, occasionally publishing his findings in the *Medical Times and Gazette*. In the
case of an outbreak in South London, he came across a set of circumstances that resembled a perfectly designed experiment. "No fewer than
three hundred thousand people of both sexes, of every age and occupation, and of every rank and station," he wrote, "were divided into two
groups without their choice, and, in most cases, without their knowledge." The single, fatal difference between those who died and those
who did not was their water supplier. The Southwark and Vauxhall water
company drew their water from the Thames, "containing the sewage of
London, and, amongst it, whatever might have come from cholera

patients," while the Lambeth water company drew theirs from a source farther upriver where the water was "quite free of such impurity."

There was absolutely no doubt in Snow's mind: cholera was spread by contaminated water. He had said it in his report "On the Mode of Communication of Cholera" in 1849 and repeated it now more forcefully in a fuller account of his theories. Sewage contaminated with cholera could spread the disease either by permeating the ground and getting into wells or by running along channels and sewers into the rivers "from which entire towns are sometimes supplied with water."

His theory, however, was dismissed. Chief Statistician William Farr led the chorus against Snow's conclusions. Farr's own study of cholera deaths revealed that the low-lying areas near the Thames—where, he reasoned, the disease mist was greatest—were most vulnerable. This risk decreased with distance from, and elevation above, the river, confirming his belief that cholera was spread through the air. In 1855, the General Board of Health reported to Sir Benjamin Hall, M.P., and Chief Commissioner of Works, dismissing Snow's evidence on the Soho outbreak. "The suddenness of the outbreak, its immediate climax, and short duration, all point to some atmospheric or other widely diffused agent still to be discovered, and forbid the assumption, in this instance, of any communication of the disease from person to person, either by infection or by contamination of water with the excretions of the sick."

They had learned nothing.

ON JANUARY 1, 1856, after six successive sewers commissions had failed to resolve the crisis, responsibility was handed over to a new body with enhanced powers: the Metropolitan Board of Works. The move was an attempt to further streamline the government of London, which as the *Times* pointed out, "had a greater number and variety of governments than even Aristotle might have studied with advantage." Bazalgette, with a growing reputation and recommendations from leading engineers of the day such as Isambard Kingdom Brunel and Robert Stephenson, soon found himself promoted to the prestigious position of chief engineer to the board. His objective was to ascertain as quickly as possible the very best design "for the complete interception of the sewage of the Metropolis."

Bazalgette, now 37, faced a daunting task. He was to transform an ancient network of ill-matching sewers into a modern system capable of serving the largest city in the world. Not only did he need to find a way to carry millions of gallons of waste away from the city, he must do so with minimum disruption to the city and its 2.5 million inhabitants. This was an engineering project so vast it could only be carried out once, and yet it must be good enough to last for hundreds of years.

He knew that despite the overall ineffectiveness of the numerous sewers commissions, they had left behind one useful legacy: the Ordnance Survey had already been asked to map the existing sewers and this proved a valuable resource. It took Bazalgette only six weeks to outline plans for the southern drainage, and an additional six to submit plans for the northern drainage. In truth, his vision for this seemingly impossible task had gradually evolved over many years and now, finally, he was in a position to put forward his plan.

"The idea was very simple," he explained years later. "The existing streams and drains all ran down to the river on both sides. All that had to be done was to carry main sewers at varying levels on each side of the river, so as to *intercept* those streams." These massive intercepting underground sewers would run parallel with the Thames on either side, three on the north and two on the south, and take the sewage east out of London. A whole network of local sewers fanning out across London would discharge into these main sewers, which could carry the waste away from the city and deposit it in the Thames on the outgoing tide. The scale of his scheme was phenomenal. In all, over 1,000 miles of street sewers and

82 miles of main intercepting sewers would stretch across 80 square miles of London.

London was built around a valley, so gravity would generally work in Bazalgette's favor: streams and drains ran down towards the river. To ensure the flow, he designed the sewers with a fall of two feet per mile. "The fall in the river bed isn't above three inches in a mile," he explained, "for sewage we want a fall of a couple of feet per mile, and that kept taking us below the level of the river." Consequently, when they reached a certain depth, great pumping stations were needed to raise sewage back to above the level of the Thames. At Abbey Mills, West Ham, in the north, a pumping station would be built to pump low-level sewage to the middle and high levels, and then on to three outfall sewers at Barking, where it would discharge into the Thames. On the south, the high-level sewers would join the low-level sewers at Deptford and run on to an outfall at Crossness, on Plumstead marshes. The beauty was in the intricacy of the plans, which combined Bazalgette's wealth of engineering experience with his astonishing ability for handling detail.

The street sewers, drawn up by Bazalgette's draftsmen to his exact specifications, resembled the shape of an egg, positioned with the stronger, rounded-end uppermost and the pointed end below. This ingenious tilted form not only provided optimum flow but meant the sewers were self-cleaning, as even when not very full there would always be water moving through at the narrow bottom end. Most important, allaying public fears that the streets of London might collapse into the sewers, the ovoid design also provided the strength required to bear the weight of the city from above.

As Bazalgette completed his plans, he felt confident that he had at last been given a mandate for action, yet he remained characteristically self-effacing in his presentation to the board. "I cannot pretend to much originality," he told them, "my endeavour has been practically to apply suggestions, originating in a large measure with others." The plans were sent first to Chief Commissioner Sir Benjamin Hall, who had ensured that he retained a supervisory role over the board as well as the power of veto, a contentious issue with board members.

It was a long wait for Bazalgette, and when the verdict finally came three weeks later the news was not good. Concerned about the stinking state of the river, Hall decided he must stick to the words of the 1855 Metropolis Act and not permit any sewage to be released into the Thames near London. Because of the lack of funds, Bazalgette's proposed outfalls were still just *within* the metropolitan boundary.

The rejection letter was a terrible blow. Even with his wife at home in Surrey awaiting the birth of their eighth child, Bazalgette once again found himself in his London office until 1:00 in the morning reexamining and reworking his proposal. Tirelessly over the next weeks and months, he prepared a number of modified plans, but none were accepted. Bazalgette was satisfied with his proposed siting of the outfalls, but in an attempt to meet the restrictions on dumping into the Thames, he suggested siting them farther downstream—warning, however, that this would be more expensive. Hall considered the idea but reported back that the government would not contribute the additional sums required. And while Bazalgette's hands remained tied, the stench over London hung like a pall, almost visible, and horribly alive with disease.

As a growing number of boards, committees, and consultants became involved in the official decision-making process, the debate continued in public. Warnings poured in from every quarter. "The waters of the Thames are swollen with the feculence of the myriads of living beings that dwell upon the banks and the waste of every manufacturer that is too foul for utilisation," declared the *Lancet*. The notion that "the abominations, the corruptions" emptied into the river would be swept away to sea was dismissed, and the prognosis was bleak. "The sea rejects the loathsome tribute and heaves it back again with every flow," the medical journal concluded grimly. For when the tide came in, all the sewage was swept upstream, and when the tide ebbed, it all came down again. "And so it kept oscillating up and down the river," said Bazalgette, "while more filth was continually adding to it, until the Thames became absolutely pestilential."

The eminent scientist Michael Faraday also contributed to the debate by conducting a simple experiment from aboard a steamboat on the Thames. He reported his findings in a letter to the *Times*. Noting that the "whole of the river was an opaque, pale brown fluid," he tore up small pieces of white card and dropped them into the water at every pier between London and Hungerford Bridge in order to test the water quality. Despite the bright weather that day, the small white squares became indistinguishable before sinking an inch. Describing the Thames as a "fermenting sewer" where "feculence rolled up in clouds so dense that they were visible at the surface," he warned that Londoners ought not be surprised if "ere many years are over, a hot season gives us sad proof of the folly of our carelessness."

Faraday was proven right. In the end it was not the medical establishment, the social commentators, or the combined efforts of Snow and Bazalgette that drove events forward. It was the weather.

The summer of 1858 was exceptionally hot and dry. By mid-June, with the temperatures soaring to record levels, the appalling odor of the Thames grew so intense that it invaded the Houses of Parliament. Red-faced M.P.s breathing in the putrid fumes ran from their committee rooms nauseated "by the force of the sheer stench," reported the *Times*. Those who could left for the country. Government slowed, speeches were left unspoken, while drapes doused in chloride of lime hung at the windows, gently waving in the warm breeze, wafting the odor through the empty rooms. Feeling the weight of responsibility on his shoulders, the designer of the new ventilation system, Mr. Goldsworthy Gurney, immediately drafted a letter to the speaker warning that he could "no longer be responsible for the health of the House."

Only a week earlier, the Metropolitan Board had made a rather canny decision. A resolution had been passed to wait until October before tackling the problem, allowing the smell to fester and grow over the long hot summer. They hoped this would finally push M.P.s to take action. They needn't have worried. On June 25, the House could stand it no longer. Owen Stanley, a Welsh M.P., led the debate, quoting medical evidence that "it would be dangerous to the lives of the jurymen, counsel and witnesses" to remain in central London at risk of "malaria and perhaps typhus fever." As panic spread, Queen Victoria's government ground to a standstill in the grip of the "Great Stink."

Not only did M.P.s find it unbearable, they feared for their lives and began to quietly find reasons not to attend their grandly Gothic palace moated by the Thames. Just one whiff of this river, they believed, would, likely as not, strike them dead. For despite two decades of research and experimentation, Dr. John Snow had failed to convince those around him that it was not smell, but water, that was a potential killer. Over a 20-year period, he had watched, powerless to save lives, as cholera returned again and again to terrorize the population, leaving over 30,000 dead. As the sun beat down on the Thames at the height of the Great Stink, and Parliament debated the most pressing issue of the day, Snow died of a stroke, aged only 45. With no one to grieve for him, and his seemingly wasted lifetime's work, all the excitement and wonder at solving the secrets of cholera's power was buried in careful plans in the debris of his desk.

Meanwhile, for all the wrong reasons, the Leader of the House, Benjamin Disraeli, did the right thing, and in a series of debates on July 22, 23, and 24, he introduced the Metropolis Local Management Amendment Bill. This finally gave the Board of Works the power and authority to proceed immediately with the new sewerage system. Underwritten by the Treasury to the tune of £3 million, at last, Bazalgette's celebrated "great work" could begin.

—

AFTER DECADES OF inactivity, the landscape of the city began to change almost overnight. Wooden workers' huts, with their tarpaulin covers, sprang up around the city, while elsewhere public walkways and parks became impassable owing to the commencement of the work. Despite the inconvenience, the project captured the imagination of the public, and this was reflected in extensive press coverage. "For good or evil, the metropolis has entered upon a work of no common magnitude," reported the *Builder* cautiously. The *Illustrated London News* was simply grateful "for the fact of the spade, shovel and the pick having at last taken the place of pens, ink and debate." The eyes of the nation were on Bazalgette.

He started north of the river with three main sewers at high, medium, and low levels, working with the natural fall of the land. Twenty to 30 feet belowground, the high-level sewer began its descent at Hampstead Heath and, over its nine-mile course, removed north London's waste until it joined the deeper midlevel central London sewer, at Stratford. Unfortunately, the beautifully simple plan met with the major obstructions of the Regents Canal and the Metropolitan Underground Railway, and fine calculations were required to negotiate this route.

The 12-mile-long northern low-level sewer posed a completely different set of problems. Beginning in Pimlico, it kept the river company as far as Blackfriars where, turning north to Tower Hill and Bow, it then made for the Essex Marshes and the pumping station at Abbey Mills. Here, eight powerful beam engines lifted the sewage over 35 feet into an outfall sewer where, together with the contents of the high- and middle-level sewers, a further five miles of pipes took the whole enormous bubbling load of north London's unwanted waste to reservoirs and the Thames, finally to lose itself in the vast English Channel. But Bazalgette had grand plans for the most troubling section of low-lying land in the center of London. Fifty-two acres were to be reclaimed altogether from the Thames, forming Chelsea and Victoria Embankments in the north and Albert Embankment in the south. This new acreage would be turned into beautiful river walkways behind high ornamental embankments, under much of which would be hidden sewage pipes, service tunnels, and an underground railway.

Bazalgette was involved at each stage, concerning himself with all the exacting minutiae of every contract. The board required endless progress reports as plans for each section of the northern, western, and southern drainage were prepared; bids scrutinized; and working methods and materials specified. Land had to be bought from unwilling owners who found themselves in the path of the sewers. Wages had to be negotiated. Every agreement generated more paperwork, and nothing was passed without Bazalgette's personal authorization. "It certainly was a very troublesome job," he admitted later. "We would sometimes spend weeks in drawing up plans and then suddenly come across some railway or canal that upset everything, and we had to begin all over again."

By 1861, it seemed that all of London was under construction. From Hampstead to Hackney and Bayswater to Balham, different sections of the sewers were at varying stages of completion. The bigger sites felt like small villages with the workforce as their temporary inhabitants. Each required its own infrastructure: railways to transport the men and their tools, storage depots for bricks and gravel, offices for the on-site engineers, and stables for the horses—trained, it was said, to leap up the banks and give their wagons a tip to turn out the gravel. Such a vast project required an army of workers. On the northern high-level sewer alone, half a million cubic yards of earth had to be excavated and 40 million bricks laid. The hours were long and men were paid for a 10-hour shift. Bricklayers earned the most at 6 shillings a day, followed by miners, who took on the tunneling for 5 shillings, sixpence, then laborers at 3 shillings, sixpence.

Bazalgette favored the "cut and cover" method, where open trenches were dug out, sewers laid within them, and then covered again with earth. This was widely accepted as the simplest and safest method. His choice of materials, however, was far more controversial. While Bazalgette always stretched himself to his fullest capacity, he could not be described as a man who took chances. His approach was characterized by attention to detail, thoroughness, and above all, persistence. Yet when he chose to use portland cement for the largest engineering scheme London had ever seen, he took an enormous risk.

The more acceptable option was Roman cement, which had been used to lay brickwork on most large engineering projects, including Isambard Kingdom Brunel's Thames Tunnel and Robert Stephenson's railways. Tried and tested, it was the obvious choice. But Bazalgette had something else in mind, because as far as he was concerned, this was unlike any project that had gone before. His brickwork would need to withstand massive pressure from above as well as constant exposure to water from within. He remembered a paper he had read some years earlier describing the construction of a small harbor on the south coast. In this project, a new material, portland cement, had been used, chosen for its special qualities: strength and durability under water.

Portland cement was made from finely ground limestone, clay, and water in exacting quantities that were crucial, as was the manufacturing process. Perfectly ground, mixed, and baked, the hardened cement appeared to be unaffected by immersion in water, and was even believed to grow stronger over time. Overheated or inexpertly mixed and it became weak and unstable. Although it had been patented only recently, in 1824, and had not been widely used, for Bazalgette its potential advantages were irresistible.

His assistant, John Grant, supervised a series of extensive tests on the new material, and after 300 experiments, Bazalgette came to have enormous faith in the product, especially in underwater conditions, where he believed it was able to withstand three times as much pressure per square inch as Roman cement. Balancing risk with caution, Bazalgette introduced a rigorous system of quality control—at a time when this was a new concept—using a machine custom-built to test every batch for strength. With no precedent available, only time would tell whether his risk would pay off in the long term. If he was wrong, flooding might occur, tunnels could collapse, perhaps even buildings.

If 1861 was characterized by progress on all fronts, the following year was marred by a series of disasters. The first, in the summer of 1862, reached the front page of the *Illustrated London News* with a report of the entire "destruction of a part of the Metropolitan railway." Due to falls in the Fleet sewer, water had built up against the wall of an embankment where the new underground railway emerged from a tunnel, causing the embankment to collapse. "A warning was given by the cracking and heaving mass," and workmen fled only moments before a massive brick wall, a hundred yards long, rose from its foundations as water forced its way underneath and slowly collapsed, taking with it scaffolding, street lamps, and pavement. Water gushed into the tunnel linking Kings Cross and Farringdon, rising rapidly to 12 feet and threatening to sweep away the old Clerkenwell pauper burial vault and "scatter its ghastly contents upon the stream." The situation was only contained due to the swift action of the contractor, who successfully opened a small outlet and diverted the floodwaters.

News of the flood spread fast as the public wondered whether their worst fears about the stability of the sewers might be realized. Bazalgette soon found himself under attack as the incident caused delays to the opening of the underground railway, but there was worse to come. The following spring an accident occurred on the site at Deptford, and this time the men were not so lucky. In the early hours of April 27, 1863, workers were excavating deep underground on a cutting for the sewer near the Deptford railway station. "Without the least warning," reported the *Times*, soil began to slip, timbers gave way, and "great masses of earth fell right down upon eight men who were buried underneath." Although five workers escaped with minor injuries, three were trapped. One, a 50-year-old by the name of Daniels, was uncovered still breathing until the soil slipped again and he died before they could reach him. It took a further four hours to reach the second man, Bray, by which time he too had perished. The final man, King, was still buried and there was no hope for his survival.

This was devastating news for Bazalgette. There was always the risk of flood, collapse, or explosion, but as chief engineer it was his responsibility to anticipate and avoid the danger and protect the men's lives. It was now almost five years since he had first taken charge of this enormous engineering project. In the early days, he had always been reluctant to take sole credit, humbly claiming a debt to many others who had gone before, but with time this had changed and he increasingly viewed the project as his own. In a letter to the *Illustrated London News* the previous year he had staked his claim, drawing attention to their mistaken belief that the plan for London's main drainage "cannot be said to be the particular design of any one man." In an uncharacteristic protest, Bazalgette had responded angrily, pointing out that of all the plans and suggestions made over the years "there is *not one* . . . which can be said to be similar to my design now in course of construction. . . . I hope you will pardon my warmth on this subject," he continued, "when I tell you that these designs have for the last ten years caused me many days and nights of anxious thought and calculation; and to assert that they are not my own is to rob me of my professional reputation."

But claiming the credit also meant accepting the responsibility, and burdened by the day's tragic events and the prospect of many difficult years ahead, Bazalgette found himself close to exhaustion. He had buckled under pressure once before, 15 years earlier, and now with so much still to be done, could not let it happen again. He pushed himself, working night after night through to the small hours in his Greek Street office, poring over papers and plans, driven by a meticulous attention to every detail and a strong sense of caution.

By the spring of 1865, he successfully tackled one of his greatest challenges: the point at which the middle sewer of the northern drainage crossed over the new Metropolitan Underground Railway near Clerkenwell. It was an extremely delicate operation, which was only resolved by constructing a subterranean aqueduct, 150 feet long, above the railway. Made of wrought-iron plates riveted together and suspended between girders 12 feet deep, the aqueduct could carry 60,000 gallons. It was initially laid five feet higher than its final resting position, then carefully lowered by hydraulic jacks into place, only inches above the engine chimneys of the trains traveling below.

By April 1865, Bazalgette was ready to celebrate the opening of the first phase of the sewer system at Crossness, where the Prince of Wales would set the colossal beam engines in motion for the first time. The southern sewer at Crossness disgorged waste directly into the Thames at high tide, but at low tide the engines were required to pump sewage 21 feet into a reservoir to await the tide. The engines were the largest in the world and had been designed for this task by the celebrated firm founded by James Watt. Each weighing 240 tons, they would be responsible for pumping thousands of times their own weight in sewage every day, first into the reservoirs and then into the Thames; but this was the day they would first be put to the test.

The Prince of Wales and other distinguished visitors came down the river by steamer to mark the opening at Crossness. They arrived at midday, and the *Fisgard* fired a royal salute. The party began their tour on the north side of the river at the Abbey Mills pumping station where the contents of the lower northern sewer were lifted to meet the middle- and high-level sewers. The Abbey Mills pumping station was not due to be officially opened until the following year but was all but complete. A striking monument to Victorian architecture, boasting a colorful exterior and elaborate wrought-iron detail inside, it had provided Bazalgette with a rare opportunity to build aboveground, and he seized it, creating a building far more beautiful than its purpose suggested.

The prince then proceeded to the south side of the river, to the pier at Crossness, which had been adorned with a crimson cloth. According to the *Illustrated London News*, every attempt was made to create a celebratory mood. Strung out along the river wall, a sea of flags caught the bright April sunlight, and pots of flowers and shrubs brought splashes of color to the terrace. A band of the Royal Marines was playing; a section of sewer was on display, lit with myriad colored lamps for ease of inspection; and Bazalgette's drawings and designs highlighted the progress of the great work. As the crowd gathered for speeches, Bazalgette spoke optimistically of his high hopes for the new drainage system. For too long, Londoners, especially the poor, had suffered sickness and mortality. This, he believed, was largely caused by insufficient drainage and, he told his audience, the new system should soon bring such deadly epidemics to an end.

The prince and his entourage were then escorted to the engine rooms. Once inside, they stood silently, momentarily awestruck by the sheer size of the massive, gleaming pumping engines and vast furnaces. Under Bazalgette's guidance, the prince turned the handle. There was a pause and a brief silence, then a tangible vibration shuddered throughout the building, as at last the enormous beams, lifting rods, and flywheels sprang into action. A deafening cheer rose from the workmen perched above in the galleries.

The day was a great triumph. At the celebration afterwards, the prince proposed a toast: "Success to the great national undertaking," he declared as all in the room raised their glasses high, "and congratulations to the eminent and skillful engineer, Mr. Bazalgette, on having made this great public work so successfully." It was a work, the prince went on, echoing Bazalgette's own sentiments on the bygone days of cholera, which would finally allow everyone to look forward to a future "when London will have become one of the healthiest cities in Europe." His claim, however, was premature.

WITH THE TRIUMPH of Crossness behind him, Bazalgette turned his attention to the Victoria, Albert, and Chelsea Embankments. Work was due to begin on the south bank in July 1866, and he was in the final week of preparation when he faced a sudden, incomprehensible crisis. That very week, on June 27, in Bromley by Bow, a laborer and his wife died of cholera.

It was unbelievable. How was the reappearance of cholera possible when so much had been done to eradicate it? For those who believed that cholera was spread by smell, it had seemed likely that the sewers would surely banish cholera from the city; now all the endless years of hard work came into question. There was utter confusion as this new outbreak began its deadly assault in the east end. With panic spreading fast across London, a breakthough came from an unexpected quarter. It was William Farr, one of John Snow's most vociferous opponents, who finally realized why Bazalgette's new drainage system had failed to save the city from cholera.

Events unfolded in the first few days of August when it became clear that the couple who had just died had lived in one of the few areas not yet covered by the northern section of the new sewage system. This in itself should not have been the cause of such a lethal problem. The East London Water Company supplied water to this area and, as their leading engineer was quick to point out in a letter to the *Times* on August 2, their water was filtered at Lea Bridge, "and not a drop of *unfiltered water* has been supplied by the company for several years." Farr, the statistician, was not satisfied. He had immediate access to the figures on cholera deaths and swiftly spotted a pattern.

A large number of deaths during this new outbreak had occurred in the six districts supplied by the East London Water Company. While he didn't yet fully understand the problem, Farr realized this could not be a coincidence; and, now, at last suspicious of water, proceeded with the same sense of urgency that had driven Snow. Farr immediately traveled to the area to investigate and soon found the irrefutable proof he needed. Two residents whose water was supplied by the East London Water Company had found eels in their water pipes. If there were eels in the water, despite the water company's soothing claims, there could be no doubt about it: the water had not been filtered.

Farr's eyes were finally opened to the very theory he had for years so categorically rejected. Within days, it became clear how the current epidemic—indeed, how all three previous epidemics—had taken hold. The couple from Bromley by Bow had flushed away infected sewage from their water closet, as so many had done before them. This sewage had passed directly into the river, in this case the River Lea at Bow Bridge, half a mile from the reservoir at Old Ford. From there it was swept upstream with the incoming tide towards the reservoir, and infected sewage was pumped into tens of thousands of homes. The residents in the infected part of the east end had been drinking water contaminated with sewage.

Farr immediately contacted Bazalgette and explained the significance of what he had discovered: that the killer disease was lurking in what should be life-giving water. Bazalgette at once made plans for a temporary pumping station to solve the problem in East London. Public notices were also posted advising those in cholera-affected areas "not to drink water which has not previously been boiled," the very advice Snow had given over a decade earlier. An inquiry was launched into the practices of the East London Water Company, which soon established that the reservoirs were not adequately protected from contamination.

For someone who had been so sure that miasma was the agent of cholera, the revelation to Farr that his former obstinacy had helped to foster previous epidemics made him champion this new truth with renewed vigor. "The theory of the east wind with cholera on its wings, assailing the East End of London, is not at all borne out," he finally admitted to Parliament. The "few cholera corpuscles floating in open air," which might have been inhaled by Londoners, were as nothing compared with "the quantities imbibed through the waters of the rivers or of ponds into which cholera dejections had found their way and been mingled with sewage by the churning tides."

With the benefit of hindsight, the shortcomings of the miasma theory were all too obvious, and it was not long before others followed Farr's lead, admitting that Snow's bold conjecture had been exactly right. In fact the key evidence had been in place in 1849 but went unrecognized by an establishment blinded by its own beliefs. Now, years after his death, Snow had at last been vindicated. Had he been taken seriously during his lifetime, many thousands of lives might have been saved—including another 5,595 in this latest epidemic.

SPURRED INTO RENEWED effort for the building of the sewers, Bazalgette now knew beyond doubt that the system would be saving the lives of countless people in the years to come. One final challenge remained: the building of the Thames embankments. On the south side, the mile-long Albert Embankment would eliminate the perpetual flooding that regularly turned Lambeth into such a quagmire. Far more ambitious were the plans for the Victoria Embankment in the north, where Bazalgette's sewers would run beneath pipes for water, gas, and—later—electricity, and alongside a new underground railway. The Victoria Embankment began at Westminster Bridge and snaked eastwards as far as Blackfriars Bridge, reclaiming over 37 acres of land from the Thames, which, now cleansed of sewage, could be celebrated in all its glory.

The *Illustrated London News* conveyed the excitement to an expectant public in regular reports. The plans for the new riverside area began to sound as though the hardworking Thames was at last to receive the breath of life. The wasted acres of mud and sewage were to be transformed into leisured walkways festooned with trees and flowers. Wide sweeping steps and seats were to be built from which to view the sparkling river and the hazy blue vistas, while exotic iron dolphins holding aloft ornamental lights would observe the scene with a timeless calm. Hidden beneath this gracious façade would lie Bazalgette's complex lines of sewers.

To create the embankments, cofferdams were constructed by driving rows of wooden piles deep into the river and plugging the gaps with a mixture of clay and earth thrown up by the excavation. The water behind the piles could then be pumped out, leaving a dry space for workers to lay the sewers. In some places bedrock was too deep to use piles, and here caissons—not unlike large iron diving bells—were lowered into the water at low tide and sunk into the riverbed to provide firm foundations. On top of the northern embankment, a new main road would be constructed, 100 feet wide with paved footpaths on either side, with a row of trees between the road and the river. This would bring great weight to bear from above. Once again, Bazalgette relied on the strength and the special water-resistant properties of portland cement. He had such faith in the material by this stage that for some sections of the sewers he used portland cement in place of brickwork, which proved cheaper as well as stronger.

By the summer of 1867, with work well under way on the embankments, the completed sections of Bazalgette's sewerage system faced their first real test. At midnight on July 26, rain fell on London with unex-

pected fury, with more than three inches in a few hours. Although Bazalgette, in his careful way, had incorporated overflow weirs into his design
to act as safety valves in the event of violent storms, even his prudent calculations had not allowed for such a record-breaking downpour. It was
hard to believe that the sewers could operate in such an unprecedented
flood.

As the sun rose, reports came in that the sewers had indeed stood up
to the forces of nature. The pumps had successfully lifted twice the volume of water they had been designed to lift and the overflow weirs had
functioned perfectly. The strain on the sewers was equaled only by the
strain upon Bazalgette himself. Although only 48, he had never had a
robust constitution, and the years of pushing himself to extremes left him
so depleted and exhausted that his health failed him for a second time.
Although the western drainage and the Victoria Embankment had yet to
be completed, suffering from nervous exhaustion, he was forced to absent
himself from the office.

THE MUCH-HERALDED OPENING of the Victoria Embankment was due
to take place on July 25, 1870. At the eleventh hour, however, the date
was moved forward to the 13th to accommodate Queen Victoria, who
would herself open the embankment named in her honor. Queen Victoria had made very few public appearances since the death of her husband
nine years earlier, but, at the urging of Prime Minister William Gladstone, was prepared to make an exception in this instance. The board had
no choice but to advance the opening in accordance with the queen's
wishes.

The change in schedule created chaos for the contractor responsible,
George Furness, who struggled to complete the work two weeks ahead of
plan. Bazalgette, too, was unprepared. Although he had ignored his doctor's advice and returned to work in just over a month, he was still weak

and in the process of recovering. Constant delays brought about by the unfinished Metropolitan District Underground Railway had already slowed work on the embankments, causing mounting frustration, and now the sudden change in plans added to Bazalgette's anxiety. In addition, dissension broke out as Furness asked his 2,000 men to work on the Sabbath, the day before the opening; the more devout members of the board objected as well. The battle was won, though, and work went ahead. Tired and worn, Bazalgette prepared his speech for the big day. If the construction workers could sweat on the Sabbath, Bazalgette decided that he, too, would make an exception for Queen Victoria.

Unfortunately, the queen herself did not take such a selfless approach. In a last-minute note to Prime Minister William Gladstone, she explained it would not be possible to attend, fearing that "already suffering much from the heat . . . it will completely exhaust her and bring on headache and neuralgia." The panic to complete had been for nothing. The responsibility fell instead to the Prince of Wales and Princess Louise, along with several other members of the royal family and numerous distinguished guests, ambassadors, M.P.s, and officials from the Board of Works. The largest and most enthusiastic contingent, however, was the general public, as thousands of Londoners poured in from all corners of the city.

From early in the day, the air was filled with the sounds of excitement and activity: the clattering of hooves and carriage wheels; salesmen hawking chestnuts and peppermint water; ballad singers, bands, and food stalls adding to the cacophony. Flags were hoisted, surmounted with floral wreaths, along both sides of the new embankment road in readiness for the royal procession. For the 10,000 ticket holders, two grandstands were decorated with striped awning and colored bunting at Hungerford Bridge and Charing Cross, which, declared the *Times*, were "chiefly occupied by ladies in brilliant toilettes and carrying bouquets still more brilliant." Others flocked to the bridges, arriving before dawn on foot or horseback to secure the best vantage points. At 10:00, as the church bells rang out, the temporary barricades at Westminster and Blackfriars were taken down, and the public crowded onto the embankments from both ends. In the rush to prepare for this moment, there had not been time to pave the great new thoroughfare, which had been hastily laid with gravel to conceal the fresh macadam beneath.

At noon sharp, the royal procession arrived in horse-drawn carriages and was joined at the Westminster entrance by members of the Board of Works, whose "ill assorted carriages," continued the *Times*, introduced "an element of the grotesque," which "rendered the whole almost ridiculous." This apparently was overlooked, and the chairman of the board rose to the occasion, graciously thanking the prince for attending the opening of what the board considered to be "the greatest public work ever undertaken in the capital of Her Majesty's Empire."

If ever Bazalgette had harbored doubts in the long years of hard work and struggle, they were dispelled now. His meticulously laid miles of pipes in the hidden world under the pavement had conquered London's unwholesome effluent, removing the poisonous tideline of filth from the city, and its millions were intent on honoring the man who had set them free from the terror of cholera. The day was a triumph for Bazalgette as he stood in the sunlight amongst the capital's most eminent men, his work acclaimed by the *Times* as "a monument of enduring fame, second to none of the great achievements that have marked the Victorian age."

As Bazalgette later acknowledged with resignation, "I get most credit for the Thames Embankment, but it wasn't anything like such a job as the drainage." With both complete, Bazalgette received the recognition he deserved. In May 1874, Queen Victoria belatedly found time for Bazalgette and he received a knighthood at Windsor Castle. In 1878, he saw his celebrated embankments illuminated by London's first electric lights, and in 1883 became president of the Institution of Civil Engineers, the highest honor in his profession. That same year, while Edwin Chadwick, still clinging to the old miasma theory, continued to lecture on the benefits of clean air, Robert Koch, a German bacteriologist, isolated the cholera bacillus and provided definitive proof that cholera was spread via fecal-contaminated water. It was clear that with the building of the new London drainage system, Bazalgette had almost certainly saved more lives than any other single Victorian.

Before he retired, Bazalgette found time to design bridges at Putney and Battersea and the beautiful suspension bridge at Hammersmith. Many new London parks benefited from his landscaping ability, with fine groups of trees making for gracious vistas and bringing a feel of the country to the center of London. As his seventieth year approached, he retired from the Metropolitan Board of Works, his fame as the man who saved London from cholera never diminishing. And in his last years at his Wimbledon home fringed by the country, a rather frail figure, with something regal about him still, he found at last the peace he had craved, in the orderly seasons, the wide summer skies, the mist in the valley, the silent midwinter. His children and grandchildren shared his days, picking apples, making hay for the two milking cows and helping in the large garden, which was his delight.

When he died in 1891, aged 72, Bazalgette could not have known that over a century later, the portland cement he had chanced his career on would still be holding the brickwork of London's sewers firmly in place. "We could repoint it," said one modern-day sewerman, "but it doesn't really need it."

—

CHAPTER 5

THE TRANSCONTINENTAL RAILROAD

It will be the work of giants . . . and Uncle Sam is
the only giant I know who can grapple the subject.

WILLIAM TECUMSEH SHERMAN, *1857*

IN THE EARLY part of the nineteenth century, the vast continent of
North America lay as it had for centuries, marked only by Native American and buffalo trails and the worn wagon tracks of those making the
journey west. Travel across the Great Plains, wide rivers, deserts, and
mountain ranges was slow and dangerous, with no guarantee of safe
arrival in California. Many making the journey from east to west preferred to travel by sea, braving a six-month voyage around South America's Cape Horn rather than attempt the hazardous overland crossing.
The only other route was to cross the Isthmus of Panama and board a
ship to San Francisco, risking yellow fever, malaria, and other deadly diseases. Thousands never reached their destination; their bleached bones
scattered across the great American desert or washed up on the shores of
South America.

At first, the railroad was not an obvious solution to these difficulties
of transportation. In 1830, when the first American-built locomotive,

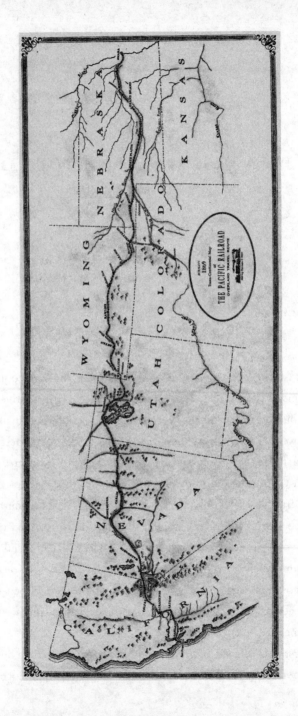

Tom Thumb, was set to compete with a horse-drawn wagon racing next to the track, its boiler burst. But from modest beginnings there followed an unprecedented boom in the railroad industry. During the 1830s and 1840s, many railroads sprang up in the eastern states, casting a filigree pattern of transport across the landscape and linking the East Coast cities. By 1850, there were over 9,000 miles of track, and it continued to be laid at a rate of over 2,000 miles a year, reaching inland to towns on the Missouri River. Despite this success, to date, no one dared attempt the daunting task of building a railroad from coast to coast.

It was not until the Civil War that that great visionary President Abraham Lincoln recognized that such a network of iron roads, carrying trains blazing across the prairie to the West, would be the answer to a desperate need: to bring the country together before its fragile unity shattered. With the onset of war between North and South on April 12, 1861, the very future of the young United States seemed under threat. Northern plans to abolish slavery had prompted seven southern states to break away and form the Confederate States of America. There was also the very real fear of losing California to the British, who were backing the South. The president needed a way to unite his divided country, to make distance disappear, to make travel to California as easy as the journey to the next town. This vast country, still in its infancy and so sparsely populated, was in danger of being torn apart.

In October 1861, as the Civil War raged across the country, a brilliant young engineer, Theodore Judah, traveled from California to Washington with a solution. A good-looking young man, his gently handsome face and equally gentle and steady gaze hid an obsession so intense that it had earned him the nickname "Crazy Judah." For years he had been surveying the hostile peaks of the Sierra Nevada range in California and had at last discovered the best route for a passenger train. He hoped to build a transcontinental railroad that would span the United States from the Atlantic to the Pacific coasts. With him he carried the blueprint for his dream: precious surveys of some of the most treacherous territory in America, forging a path across miles of uncharted and hostile land. At a time when many had never seen a passenger train, it was a plan only a visionary or a madman could have pursued. While in the past, Congress had considered the idea of a transcontinental railroad as absurd, like trying "to build a railway to the moon," Judah himself entertained no such doubts. "It's going to be built and I'm going to have something to do with it," he promised his wife, Anna.

The greatest obstacle was the Sierra Nevada, a massive, immovable barrier between the Pacific Coast and the rest of America, stretching 400 miles from north to south and 80 miles wide. Winter arrived in early autumn in the mountains; the passes and valleys would fill up with snow, sometimes to a depth of 50 feet. Spring brought snowmelt, causing flash floods and torrents choking the river valleys. Before any railroad from California could reach the wide-open spaces of the great American desert, it would first have to cross this forbidding fortress with its treacherous peaks towering 14,000 feet, molded from the toughest granite and descending from dizzying 45-degree angles into deep ravines and valleys. No one had ever laid a railroad at such heights, yet Judah was convinced he had mapped in his surveys of the Sierras a way for this to be done.

The young engineer had an impressive reputation. He had designed the innovative Niagara Gorge Railroad—also said to be an impossible task—when he was only 25, and later built the Sacramento Valley Railroad, California's first track. Given the short history of railroad engineering (it had been barely 30 years since George Stephenson's famous Rocket, the first passenger locomotive between Liverpool and Manchester), this had made him the expert in the field. In the process of compiling his survey, Judah had made 23 trips into the Sierras in search of the perfect route. Moving laboriously over ridges and through canyons on horseback, he had measured endless angles and inclines in the hopes of finding a suitable path for a train. His loyal wife, Anna, rode at his side, making delicate pencil drawings of the awesome landscape while her husband worked. After years of study, he believed he had found a route through at 7,000 feet, but it would take them to one of the most dreaded parts of the mountain range: the Donner Pass.

In 1846, a group of pioneers led by Jacob Donner had set out across the Sierras and reached this pass, the final stage of their journey west, when snow began to fall heavily. This signaled the start of what was to be a bitter winter, during which snow fell without respite. The pioneers endured winds that cut like a knife, blowing down from the icy peaks, and temperatures that plummeted to minus 40 degrees. Soon food supplies ran out and they were trapped in a vast white universe, while the chances of survival became desperately slim. But five months later when they were found, a few of the toughest had survived. It was rumored the price of survival was cannibalism; men and women had been forced to eat their own dead family members. This desolate spot became known as the Donner Pass.

When Judah arrived at the Donner Pass, he could see at once that if his railroad reached this far, beyond was a natural pathway through the mountains, along the Truckee River as it wound its way gently downwards through the canyon towards the plains of Nevada. He went at once to Washington, D.C., to distribute his findings to every senator and representative in Congress and, most importantly, to the president himself. The persuasive young engineer clearly made a significant impression. He seemed to have all the answers and could anticipate every question; his youthful face came alive as he displayed his expertise. "His knowledge of his subject was so thorough," wrote one representative, "his manners so gentle, his conversation on the subject so entertaining, that few resisted his appeals." The railroad, Judah argued passionately, would bind the nation together and create incalculable commercial potential. "Its profitableness will exceed that of any known road in the world," he declared. As for the Civil War, which preoccupied President Lincoln above all else, a railroad would provide invaluable transport for troops and ammunition across the vast country.

In President Lincoln, Judah had finally found the man with both the authority and the vision to help him make his dream a reality. Lincoln believed a transcontinental railroad was "imperatively demanded in the interests of the whole country," and put his signature to the Pacific Railroad Act of 1862. In tribute to Judah's "indefatigable exertions," senators and congressmen alike signed a grateful letter to the engineer, praising his explorations in the Sierra Nevada which, they felt, had "enabled many members to vote confidently on the great measure."

According to the terms of Lincoln's Pacific Railroad Act, two companies were authorized to take up the challenge, making them rivals and possible enemies right from the start. The Central Pacific Railroad Company, already assembled by Judah in Sacramento, California, would start working its way east across the Sierras. At the same time, a second company, the Union Pacific Railroad Company, was authorized to move westwards from the Missouri River, on the border between Iowa and Nebraska, where the existing railroad network ended. Somewhere in the unmapped wilderness in between, the two companies would meet.

With generous government grants awarded for each completed mile of railroad, the competition to "make tracks" would be fierce. Though the two companies would be paid for each mile of track, it was to be on a sliding scale according to the difficulty of the terrain. For track laid on flat ground, the rate was $16,000 per mile; for steeper land, this would rise to $32,000, reaching a staggering $48,000 per mile for track laid through the mountains. The Pacific Railroad Act also granted both companies a 400-foot right-of-way along the most direct route and 6,400 acres of public land alongside the railway to sell or retain for each completed mile of track. Once the railroad was completed, this would become some of the most valuable land in the country, and the company that moved fastest could claim the greater chunk of this prized real estate. As if any further incentive were needed, at a later stage, the government doubled the size of this initial land grant, taking it up to 12,800 acres per mile of track. This was the beckoning—and elusive—goal that drove both companies on. Cities would rise on the empty plains; millions would come and millions would settle. It was the race of the century, and while Judah and Lincoln may have been guided by loftier principles, others were motivated by only one thing: money.

Building the railway from California in the West through the massive mountain range would clearly be a vast and complex undertaking requiring machinery, men, funds, and skillful engineering. No major building company or contractor stepped forward to take up the challenge. It was left to four shopkeepers from Sacramento, California, who, with a heady confidence born from selling hardware and supplies, joined together to answer the president's challenge. The "Big Four," as they became known, decided they had as much know-how as was needed and proceeded one sunny January day in 1863 with the inauguration ceremony of their railroad: the Central Pacific Railroad Company.

The citizens of Sacramento turned out in disbelief to witness this momentous event. The local paper had paid for a band, whose members played their modest repertoire until Charles Crocker, one of the four shopkeepers, rising to the glorious moment, stepped forward to reassure all within earshot of their fine purpose. Crocker ran a successful dry goods store in Sacramento and was no stranger to risk taking. The uneducated entrepreneur had left his rural Indiana home at an early age and come west in the Gold Rush, only to find he could make more money supplying miners from a series of small stores.

"We all said to each other that if anyone could do it, we could," he announced to the incredulous crowd with his usual confidence. He assured the assembled company that, even as they stood, work was going forward—the work that would transform the country, that would be the inspiration for the world, that would put Sacramento on the map. A priest was then found to say a prayer.

After introducing his partner, the president of Central Pacific, Crocker stood back and let Governor Leland Stanford take the stage. "Fellow citizens," Stanford began grandly, "we may now look forward with confidence to the day, not far distant, when the Pacific will be bound to the Atlantic by iron bonds that shall consolidate and strengthen the ties of nationality and advance with great strides the prosperity of our state and of our country." After Stanford had struck the right inspiring note, it was time for the highlight of the occasion, a carefully planned symbolic gesture to mark the beginning of the Transcontinental Railroad. On cue, a waiting wagon was pushed through the mud while Stanford, to mark the gravity of the occasion, added a shovelful of dirt to the already plentiful supply underfoot. Crocker asked the thinning crowd for nine cheers instead of the usual three to underline the significance of the occasion.

If there was a sense of cynicism among the crowd, it was with good reason. The four shopkeepers charged with creating half of America's great iron road had no experience whatsoever of railroads. In fact, the Big Four were not blessed with any knowledge of engineering or construction at all. Along with Collis Huntington and Mark Hopkins, Crocker and Stanford were the men recruited by "Crazy Judah" to build his railroad. Stanford, as well as being a successful local politician, ran a wholesale grocery business, while Huntington and his partner Hopkins owned a prosperous Sacramento hardware store. All had left the Northeast during the Gold Rush and made their fortunes providing supplies to the prospectors and settlers. They were ambitious to the point of folly, and when Judah had approached them as possible investors three years earlier, they had quickly seen the potential of his proposal.

Reporting on the opening the following day, the *Sacramento Union* made little attempt to conceal its lack of faith: "Underlying all the enthusiasm there was a fear that it was a farce and not a fact which was being inaugurated." Yet this was America, the land of opportunity, and the Big Four themselves were untroubled by self-doubt. "None of us has any experience of building a railroad or of running one," Crocker happily conceded, "but we know how to run a business. The people of this state are crying out for a railroad," he continued, "and by God, sir, we mean to give them a railroad."

But as glasses were raised to the glorious future of the Central Pacific Railroad, the towering Sierra Nevada cast a long shadow over the Big Four's seemingly foolhardy optimism. Even with the best will in the world it was near impossible to see how four local shopkeepers could overcome such an awe-inspiring obstacle. Although the first 20 miles out of Sacramento were relatively flat, the Central Pacific would then come up against the soaring ridges and plunging ravines of the Sierras. Even Judah acknowledged that "rivers run through gorges or canyons in many places from 1,000 to 2,000 feet in depth, with side slopes varying from perpendicular to an angle of forty-five degrees." The ridges were in many places so narrow at the summit "as to leave barely room for a wagon road to be made without excavating the surface of the ridge."

It was widely believed that the maximum gradient for a train was 2 feet over a distance of 100 feet. Using just a theodolite and poles for taking measurements from intersecting straight lines, Judah had painstakingly calculated every wind and curve of the landscape to forge a route through. He spoke optimistically of "a remarkable regularity of surface" in the Sierras, and found a twisting route that would meet the required grade of only 105 feet per mile. Where the gradient was impossibly steep, Judah planned to tunnel through the obstinate granite rock face of the mountains. Fifteen tunnels were required, with the longest, the Summit Tunnel, not far from the Donner Pass, stretching 1,659 feet.

The sheer-sided valleys in the Sierras would have to be crossed with bridges, often hovering hundreds of feet from the ground. Where the chasms between the mountains were too wide to accommodate a bridge, vast trestles were needed, the first of which would be constructed on the approach to the American River just outside the town. At higher levels, the unpredictable weather would also pose a huge threat and to face it, Judah anticipated the use of plows to clear a way through deep snowfalls. But he miscalculated the depth of snowfall in the Sierras to be not much above 6 feet—a severe underestimate, since depths of between 40 and 60 feet had been known to fall.

Quite apart from the engineering challenges involved in tackling the Sierras, the Big Four also faced the difficulties of financing the project. In order to make a start, both companies needed funds in advance to buy materials and stock and to pay a workforce. They were eligible for a loan from the government in the form of bonds, but the companies first had to find investors to buy these bonds in order to generate the necessary cash. Moreover, any monies would be released to the railroad companies only after the first 40 miles. For the first 40 miles of track they had to raise their own finance, and this was proving difficult.

"Don't you have anything to do with those men, Stanford, Hopkins, and Huntingdon," one San Francisco banker is alleged to have said. "Don't you put any money into their schemes. . . . Nobody in the world could get that road through." To add to their difficulties, many local businessmen—owners of shipping, freight, and stagecoach companies—stood to lose trade from the railroad, and did all they could to block its success. Thus, the Big Four borrowed extensively, often using their own houses and personal assets as collateral to finance the railroad. While Crocker fought to get work under way, Huntington traveled to New York on the trail of investors. Although he did eventually manage to find some to fund at least the first 40 miles, as security he had to offer everything the Big Four owned among them. They now stood to lose everything.

To make headway, the Big Four resorted to a scam, which would prove to be the first of many and would serve to drive a wedge between them and their rather more honorable chief engineer, Theodore Judah. Although it was plain to see that the ground around Sacramento lay relatively flat for as far as 20 miles outside the town, Stanford hired a team of geologists willing to testify that the mountains began only seven miles outside Sacramento, then sent the report to the president's representative in Washington. To his delight, this report was duly accepted by the president, which entitled the shopkeepers to the highest track rate of $48,000 from only seven miles outside of town! Although the Big Four stayed just on the right side of the law, Judah was becoming increasingly uneasy about the methods they were prepared to use to obtain funds.

Crocker, meanwhile, had resigned from the board of directors before work commenced to form his own contracting company, and soon won the lucrative contract to build the first 18 miles of track. But he faced major problems of supply. His team could cut wood for the cross ties from local forests but everything else had to come at great expense from the East by boat around Cape Horn. Bad weather conditions around the Cape meant insurance prices soared. And the workforce, not oblivious to the fact easy money might be made, went on strike. A man, a horse, and a cart, they decided, was worth $5 a day, not $4. A general air of inefficiency hung over the work site. After subcontracting the work to smaller firms, 200 men had begun grading to level the ground, but it wasn't long before they discovered that a few feet below the soft topsoil in the foothills lay a mixture of rubble and sand they called "concrete." Heaps of topsoil littered the workplace in disorganized array, and as soon as workers hit the concrete, they found reasons to leave the site and move on to easier or more lucrative jobs. Clearly there was a need for a strong foreman.

As the pressure increased, the Big Four's business partnership with Judah deteriorated. With his noble vision of uniting a continent, Judah came to loathe what he saw as the shopkeepers' despicable money-making schemes. "I cannot make these men . . . appreciate the 'elephant' they have on their shoulders," he told his wife despairingly. "They won't do what I want and must do. We shall just as sure have trouble in Congress as the sun rises in the East if they go on in this way. . . ." By July 1863, Judah was so distressed that he was ready to quit. But he could never really consider abandoning his grand scheme, so he decided instead to abandon the Big Four. He aimed to buy them out; but to do so, he, too, needed investors.

After contacting prospective financiers in Boston and New York, Judah set out for the East Coast in October 1863, taking a shortcut through the dangerous Isthmus of Panama in his eagerness. The initial response had been encouraging, and Judah was in good spirits as he wrote to a friend en route of his "relief in being away from the scenes of contention and strife" and of his hopes for the future. "There will be a radical change in the management of the Pacific railroad," he predicted optimistically. "It will pass into the hands of men of experience and capital."

The identity of Judah's new backers would remain a mystery, and how close he came to wresting control from the Big Four would never be known. Boarding the steamer for New York in the pouring rain, Judah became drenched while helping women and children aboard and shielding them from the downpour. "He could not see them exposed to the rain and not try to do his part, and more, for women and children who had no one to help them," recorded his wife, Anna. "I feared for him and remonstrated for I knew he was doing too much, but he replied, 'Why I must, even as I would have someone do for you—it's only humanity.'" But, exhausted, Judah had contracted yellow fever in the tropical downpour. His wife was helpless as, on the eight-day journey, she watched her husband submit to the inevitable succession of terrifying symptoms. In the cramped conditions onboard, and with no effective medical help, he died, aged only 37, his great dream of heading a more altruistically minded company going with him to the grave.

Judah died on November 2, 1863, but this did not prevent the Central

Pacific Railroad Company from celebrating its new star attraction a week later on November 9: a massive wood-burning locomotive 50 feet long and twice as high as the tallest man in Sacramento. They christened it the Governor Stanford. It was a powerful beast, and where possible it had been painted in rainbow colors: Chinese red, lantern orange, forest green. In the gray chill of early November, the citizens of Sacramento watched as this highly touted asset of the company was paraded along the two miles of existing track, which ended just outside the town. All available space was taken by every would-be young train driver in the West as it puffed along the track, its whistle blowing and flags flying.

Despite the appearance of success that the new engine brought the company, the reality was very different. By the end of 1863, the Big Four were on the verge of bankruptcy and facing the future without their brilliant chief engineer. Crazy Judah's dream had brought them this far but now they were on their own. Even the ebullient Crocker was close to defeat. "I owed everybody that would trust me," he said despairingly, "and would have been very glad to have had them forgive me my debts, and take everything I had, even to the furniture of my family, and to have gone into the world and have started anew." With crippling debts growing by the day, a very big question mark hung over the entire venture.

JUST ONE MONTH after the train named in his honor had made its first outing, Governor Leland Stanford received news of another celebration over 1,660 miles away. The telegram that arrived on December 2, 1863, was like a starter's pistol. That day, on the banks of the Missouri River, the Big Four's rivals in the East were celebrating their own launch. Stanford immediately drafted a carefully worded reply. It was addressed to Dr. Thomas Durant, the new leader of the Union Pacific Railroad Company. On the face of it, it was a message of encouragement, but little genuine warmth was wasted on the rival company. "California acknowledges with joy the greetings of her sister, Nebraska, and will prove her fraternal regard by her efforts to excel her sister in the rapidity with which, carrying the iron bands of Union, she seeks a sisterly embrace. Mountain and desert shall soon be overcome." The race had truly begun.

It soon became clear, however, that the new vice president of the Union Pacific, Dr. Durant, was in no hurry to lay track—although he was far from idle. If the Big Four were guilty of using legal loopholes wherever they could, Durant was in an entirely different league. Where the Big Four were wily, Durant was completely without scruples. He was tall and dark, with thoughtful eyes that seemed to be forever weighing up the situation at hand and quietly calculating profit and loss. Cutting a flamboyant figure on the New York business scene, he dressed exquisitely, expensively camouflaging himself as a gentleman and glorying in his reputation as a man who "sees no obstacles, fears no difficulties, and laughs at impossibilities."

Born to a wealthy merchant family in Massachusetts, Durant had studied medicine but had quickly lost interest. Railroads promised far richer rewards, and while working for the Chicago and Rock Island Railroad and Mississippi and Missouri Railroad in the 1850s, he had rapidly established himself as a shrewd and devious operator. Not for Durant the grand ideals of nation building or the challenge of construction; only the alluring prospect of picking the bone clean at the great fat feast of the railway bonanza spurred him on.

While the Central Pacific laid track, Durant laid plans. His careful preparations had begun well in advance of the Union Pacific's opening ceremony. To get the railroad started, at least 2,000 shares had to be sold in the Union Pacific Railroad Company, but no single person was allowed to control more than 200. This was a setback for Durant, who wanted the lion's share of the power for himself. But for every problem, Durant had a solution—often an illegal one. He had set his plans early on, making gifts to well-chosen friends and associates, men like himself who could appreciate the bends and curves in the road to wealth—men with useful political connections. He wooed them with extravagant parties, and lavished attention on them until, one by one, congressmen and leading businessmen alike fell into his trap. He would lend his friends the money to purchase shares in their names and then buy the stock back from them almost immediately. And so it was that Durant alone came to control 1,000 of the 2,177 shares sold, giving him complete control over the Union Pacific.

Durant could now bring his plan to fruition. With the support of his business associates, he discreetly set up another company, with himself as president: the Credit Mobilier. Construction deals would go out through Credit Mobilier, which would then charge exorbitant rates from the Union Pacific, which in turn would claim money from the government. Many of the major stockholders in Credit Mobilier, including Durant, could also be found on the Union Pacific list of principal investors and directors; and with no one any the wiser, as one company grew more successful, so did the other.

Unusual for Durant, he chose a man of integrity as his chief engineer, Peter Dey. Durant had met Dey while working for the Mississippi and Missouri Railroad. Dey worked hard for the Union Pacific, but as time passed he became increasingly suspicious of Durant's motives. The first stumbling block was the continual indecision over where exactly along the Missouri River the railroad should begin. For months the proposed starting point changed repeatedly, and each time Dey was asked to prepare new maps. One week Durant would announce that the railroad would connect to the existing Mississippi and Missouri line, the next he would be talking up the idea of connecting with the Galena and Chicago Union Line. "It is difficult to make surveys," Dey wired Durant despairingly in January 1864, "without forming some idea of what you are doing and what it is for."

Unknown to Dey, Durant was buying up stock in one line while it was cheap, then secretly selling it off at a tidy profit once the idea was quelled that this would be the connecting line. Meanwhile, he also bought up stock in another line at a rock-bottom price before announcing that this line would in fact be the connecting line, at which point the stock would soar in value and Durant would cash in once again. With each transaction, Durant pocketed a small fortune—and all before a single line of rail track was laid.

As the year wore on, Dey's frustrations increased. "Mr. Durant has got the whole thing in his hands," he wrote, "but he is managing it as he does everything else—a good deal spread and a good deal do-nothing. I have been so disgusted with his wild ideas that I have been disposed repeatedly to abandon the whole thing." To his dismay, he became suspicious that Durant was involved in yet another underhand scheme. He realized that through Credit Mobilier, Durant was paying his construction company $50,000 for each mile of track, while Dey knew from his own surveys that it cost only $30,000 at most. Over a hundred miles of track the discrepancy was enormous, and while Dey didn't know where the money was going, he was under no illusion that the extra $20,000 per mile was somehow winding up in Durant's pocket. It was the last straw for Dey, a man of recognized ability and probity, and he wanted no part in it. "No man can call $50,000 per mile . . . anything else but a big swindle," he declared. He tendered his resignation in December 1864.

April 12, 1865, brought the end of the Civil War and, two days later, the assassination of President Lincoln. While the nation grieved, Durant, oblivious to the tragedy, was on a charm offensive. The object of his attentions was the man he hoped to recruit as his next chief engineer, the celebrated war hero General Grenville Dodge.

Grenville Dodge was a natural leader who had risen from a modest background as the son of a laborer to a distinguished military career, and brought with him an authority and respect for discipline from his long years in the army. He was a man to rely on, as decent and honest as

Durant was not; integrity and honor were built into his very bones. During the Civil War, he had left an impressive trail of military railroads behind him, which had won him supporters in high places, including President Lincoln himself, and this explained much of his appeal to Durant. With his rather piercing eyes, Dodge could usually see straight through anything underhanded, and he was under no illusions about Durant, confiding to his brother that he was only too aware that the doctor might be trouble. Consequently, although he did eventually accept Durant's offer, it was on condition that he have absolute control.

So with General Grenville Dodge at the helm, on May 1, 1866, the Union Pacific at last began to move swiftly westward. Dodge asserted himself immediately by making a final decision on the starting point for the railroad, favoring a route from Omaha on the Missouri, across the broad Platte Valley, running 600 miles west through Nebraska into Wyoming, towards the Rockies. The Platte Valley route was one that the buffalo had taken for centuries and their hooves had left behind a well-worn trail. Although it was not the most direct route, it was the easiest and Dodge knew this land well. He owned one of the last houses in the frontier town of Council Bluffs, Iowa—literally poised on the edge of the wilderness—the very point at which he thought the railway should begin. But the buffalo trail also had its drawbacks: it led to an unmapped wilderness; it was, according to contemporary accounts, a "terrible country, the stillness, wildness and desolation of which is awful." There was little or nothing in the way of natural resources, like timber. And the only people inhabiting this land were the Native Americans, whose reputation terrified the workers.

For the first 150 miles, Dodge's men would be laying track through Pawnee land, thought to be a relatively friendly tribe; but thereafter, the Union Pacific would face the Sioux, Cheyenne, and Arapaho tribes, who had suffered repeatedly at the hands of the white man and were not so acquiescent. Dodge had personal experience of their determination and methods of defense. In June 1865, he had witnessed an attack by 1,000 Sioux, Northern Cheyenne, and Arapaho. Soldiers and civilians had been killed in the most brutal way, including one lieutenant who "was horribly mutilated; his hands and feet were cut off and his heart torn out. He was scalped and had over 100 arrows in him."

The answer to the growing hostilities came in the form of the Casement brothers, described by one associate as "the biggest little men you ever saw—about as large as twelve-year-old boys, but requiring large hats." Jack, the older of the two, was 5 feet 4 in his boots and sported a Cossack-style hat, with Russian Romanov looks to match; he carried a bullwhip at all times and an ivory-handled Colt in his belt. He did not bother with charm, preferring it to be known that he was tough and that his toughness was his greatest asset. He expected unquestioning loyalty and usually got it. His younger brother, Dan, who was even shorter, was a slightly softer version of Jack, but not by much. In the spring of 1866, they signed on to take over construction of the railroad. Their reputations being far bigger than their stature, rumor had it that they took the job only in the hopes of clashing with the Sioux and the Arapaho. With the Casement brothers under the command of Dodge, the Union Pacific seemed unstoppable. Before their arrival, the railroad had struggled to make the 40-mile mark, with track being laid at the rate of half a mile per day. A year later, the pace had rapidly increased to three miles per day.

The surveyors took the lead, working hundreds of miles ahead, exploring the terrain, mapping the details of the route, and calculating the necessary cut and excavation to make a level for the railroad. As they worked, so-called tie-hacks went west in search of forests, as the prairies offered nothing in the way of natural resources. When the weather was on their side and the timber plentiful, the best of these men could hand-cut 60 ties a day. Behind the surveyors came the graders, who followed their instructions to level the ground. Despite the overall regularity of the surface of the Platte Valley, there were still ridges and hills that needed leveling. One thousand men working in teams set about grading and leveling the ground like a factory production line.

Fast on the heels of the surveyors and graders were the tracklayers themselves. Laboring under the harsh summer sun, in their blue uniforms and heavy leather boots, the men made an impressive sight. Dodge had the work organized with military precision. "Thirty seconds to a rail for each gang, and so four rails go down a minute," observed one journalist. "Close behind the first gang come the gaugers, spikers and bolters, and a lively time they make of it. It is a grand Anvil Chorus that those sturdy sledges are playing across the plains. It is in triple time, three strokes to a spike." Jack Casement moved between the men, commanding and cajoling, often offering bonuses and fresh tobacco for faster work.

A growing number of men joined the Union Pacific, a mix of Irish, Blacks, Mexicans, Scots, and Civil War veterans—some of whom had been fighting each other only a few years before. As the labor force reached 10,000, the company needed a small army of cooks, supply men, and guards to keep up with the workers as they made their way across the plains. The main food supply was the 500 cattle that trotted alongside them, and most of the men were content with a diet of beef, bread, coffee, and spirits when available. For sleeping, long, 50-foot trucks were rigged with bunks, but many of the men preferred to bed down in the open to avoid the fleas and vermin that were constant companions in the cramped conditions of the trucks. Those sleeping in the open faced the legendary Platte storms that came thundering across the plains with breathtaking fury, white lightning, and beating rain. But another horror, far more terrifying than the storms, preyed on the minds of these men. Every worker slept with his gun beside him, which provided his only protection in that vast, lonely wilderness against the real and imagined horrors of the silent killer, which came with gun, knife, and arrow to commit unspeakable ritual murder.

The first encounter with the Sioux came in the summer of 1866 and left the men feeling fearful. A large party of warriors, led by the notorious Chief Spotted Tail, approached the camp, curious about the strange activities of the white men. They were shown around, and expressed great interest in a locomotive that stood on the track. A race was suggested to be run between the Sioux horsemen and the train. The riders were quickly overtaken by the powerful engine, and when the guard blew the whistle, the astonished warriors swerved abruptly from the track, flinging themselves to the sides of their horses in a defensive maneuver. The Sioux, who took great pride in their horsemanship, had been beaten by the great iron horse. For the braves, this was an unparalleled psychological defeat, causing deep humiliation. After sharing an uneasy lunch together, Chief Spotted Tail demanded that his men be allowed to take with them as much food as they could carry, and the mood soured further. When the foreman refused the demand, the chief was enraged, threatening to attack by night with hundreds of warriors. "That was the last we saw of them—they had seen our guns," wrote the foreman, who had been careful to show off the men's arsenal. "However, fearing they might return, we doubled our night patrol for some weeks."

Life on the railroad was tough, with the men often working 14 hours a day, 7 days a week. At the first opportunity after payday, a trip was quickly organized to the nearest settlement that had a saloon supplying unlimited alcohol, and some rare—and invariably expensive—female companionship. These settlements, which followed the railroad, caused almost as many problems as the Indians themselves. True dens of iniquity devoted to drinking, gambling, whoring, and shooting sprees, these "Hell on Wheels"—as they became known—exploded into existence in the winter of 1866–1867. The first, but by no means the worst, was 290 miles from the starting point of Omaha, at North Platte.

"Every gambler in the Union seems to have steered a course for North Platte," wrote Henry Morton Stanley, a young journalist trailing the Union Pacific, who would later win fame tracking down Dr. David

The leading scientist Michael Faraday, appalled by the stench of the river, presents his card to Father Thames. *Punch* 1855.

FARADAY GIVING HIS CARD TO FATHER THAMES;
And we hope the Dirty Fellow will consult the learned Professor.

"Toshers" were named for the "tosh," or scraps of copper, they hoped to recover from the waste in the sewers.

1860 cartoon showing a water pump spreading disease.

Joseph Bazalgette, chief engineer to the
Metropolitan Board of Works, inspecting
the works at the Northern outfall sewer.

Joseph Bazalgette, circa 1880.

The Victoria Embankment with the sewers, underground railway, and
service pipes under construction from the *Illustrated London News*, 1867.

General Grenville Dodge, Civil War hero and chief engineer of the Union Pacific Railroad from 1866.

Charles Crocker headed the construction company that built the Central Pacific Railroad.

Dr. Thomas Durant, the corrupt vice president of the Union Pacific Railroad.

Visionary Theodore Judah was called "Crazy Judah" for his obsession with building a transcontinental railroad.

The Central Pacific Railroad used snowplows to continue the work through the Sierra Nevada.

An early poster advertising the opening of the Transcontinental Railroad in 1869.

The opening ceremony at Promontory Summit, Utah, on May 10, 1869, when the last spike was driven, joining the Central Pacific and the Union Pacific Railroads.

Vicomte Ferdinand de Lesseps,
hero of the Suez Canal, was
destroyed leading the
French effort to build the
Panama Canal in the 1880s.

The charismatic John Stevens, who
suddenly resigned as chief engineer in 1907.

The treacherous Culebra Cut, the lowest
point through the mountains of Panama,
where so many lost their lives.

Testing the great locks at Gatun. These were large enough to accommodate a ship the size of the *Titanic,* with room to spare.

The opening of the Panama Canal, in August 1914, was eclipsed as World War I broke out in Europe.

Chief engineer Frank Crowe in 1932. He became known as "Hurry up Crowe" as he completed the dam well ahead of schedule, collecting a bonus of between $250,000 and $350,000.

High scalers risked their lives swinging out over the canyon walls, suspended hundreds of feet above the Colorado River. They used explosives to remove loose rock and smooth the rough surface of the canyon walls.

The Hoover Dam under construction in Black Canyon in 1933. The concrete was poured in a series of blocks, then allowed to cool before more concrete was added.

The Hoover Dam nearing completion in 1935, showing the intake towers and the lake beginning to fill up behind the dam.

Livingstone in Africa. "Every house is a saloon, and every saloon is a gambling den. Revolvers are in great requisition . . . hundreds of bull whackers walk about, and turn the town into a perfect Babel." Even compared with the wild days of the Gold Rush towns, Stanley added, "this town outstrips all yet." While the money lasted, a man could drink himself into oblivion and lose his wages or his life equally easily in these uproarious caverns devoted to fulfilling every need of the flesh and operating where law did not. With men still drunk from the night before, and often demoralized after gambling away the week's wages, Jack Casement had to work double-time to keep the workers in line. As winter closed in, his problems were compounded by raging blizzards and snow drifts, which often halted work and left men more time to spend in the saloon.

Over 300 miles of track had been completed by the spring of 1867, but severe flooding followed and work came to a standstill once more. From New York, Durant declared he was unhappy with the recent delays and spiraling costs, and so, with his entourage and hangers-on, the Wall Street wizard left for the outback determined to apply his own personal alchemy for turning dirt into gold. From the day he arrived, he clashed with Dodge, whom he accused of overspending on labor and materials. He then announced that he intended not only to employ his own engineers but to substitute the route planned by Dodge by a much *longer* one—on which he could secretly make even more money. Dodge was outraged. He was increasingly aware that Durant's purpose in building the railroad was to create his own private fortune. It was only when he threatened to quit that Durant, under pressure from some of his associates, was forced to back down, and Dodge had his way.

But with each mile west, Dodge and his men moved deeper into the heart of dangerous Native American territory, and in the months that followed, a series of savage attacks rekindled an ever-present fear among the railroad workers. Trains were derailed, track ripped from its moorings, and workers slaughtered by small bands of warriors. On a single day in May 1867, a team of six were attacked and scalped in the plains of Nebraska; and four graders, working a hundred miles ahead of the track layers, were brutally killed. The Sioux attacked everyone with demonic fury: settlers, ranchers, and stagecoach passengers; men working in isolated groups along the line were especially vulnerable. It was a will-o'-the-wisp enemy, creeping up in silence, blazing into sudden attack like demons from hell, to leave mutilated victims behind.

In June, Dodge lost one of his most experienced engineers to another tribe, the Arapaho, when the unfortunate man became separated from his group. He was found several hours later, his body pierced by 19 arrows. His death was followed a week later by the savage killing of Dodge's finest surveyors. "Trouble never comes single," the general confided to a friend, all too aware of the effect these deaths were having on morale. "Indians on the plains have been very bad for two weeks. They have been attacking everything and everybody. . . . It takes the nerve out of them losing so many. . . ."

As the dangers increased, so did the recklessness of the workers. During the summer, the Union Pacific pushed forward to Julesberg, Nebraska, a town that rapidly acquired the reputation as "the wickedest in America." Journalist Henry Morton Stanley described North Platte as a "quaint frontier hamlet" compared with this new center of "debauchery and dissipation." The women appeared to be the most reckless, he reported after visiting the King of the Hills dance hall in Julesberg, "and come in for a large share of the money wasted. . . . gliding through the sandy streets in Black Crook dresses, carrying fancy derringers slung to their waists, with which tools they are dangerously expert." It was a town, he continued, where men "would readily murder a fellow creature for five dollars," and "not a day passes but a dead body is found somewhere in the vicinity."

These casually violent, alcohol-fueled towns gave birth to America's Wild West, making legends of the desperadoes and gunfighters who ruled them. And Julesberg represented the full-blown fury of the Wild West; its shimmering saloons promising everything through a drunken and befuddled haze. Loyalty, truth, and integrity in such a place were no protection. The law that settled every question was the gun. For General Dodge, a man of absolute integrity, Julesberg posed another problem: already angered by the "profanity, vulgarity and indecency" he witnessed there, he was further enraged by the fact that this was taking place on land that belonged to the Union Pacific. So when a gang of gamblers seized plots and refused to pay the Union Pacific land agent his due, Dodge summoned Jack Casement. "Go and clean the town out," he ordered.

This, Dodge conceded, "was fun for Casement," who rode into town with 200 armed railwaymen and ordered them to "fire at will." Reports of what came to be known as the Julesberg Massacre claimed that 30 men were killed that day and that many more were injured. Before leaving the town, Casement proudly took Dodge on a visit to the local cemetery. "They all died in their boots," he said, pointing to the rows of freshly dug graves, "and Julesberg has been quiet ever since." But while Dodge believed the Union Pacific's swift action had taught the gunslingers and gamblers a "lesson that lasted for some time," it would only be a year before the next massacre, whose significance would quietly gather force in the wild country and lead to an inhuman horror that memory could never erase.

WHILE THE UNION Pacific was racing across the prairies, the Big Four were at a standstill in the Sierras. Leland Stanford, with his political influence, did have some success in securing extra state aid, and Collis Huntington successfully organized supplies to come from the East. Charlie Crocker, however, as general superintendent of the company and in charge of construction, was finding it almost impossible to retain workers. Events reached the crisis point in 1865, when the railroad reached Colfax, a stagecoach ride from the gold and silver mines of Nevada. Temptation proved too strong to resist, and of the 2,000 men Crocker had managed to assemble, 1,900 deserted for the mines. In desperation, Crocker found himself paying huge sums of money to ship in newly arrived Irish immigrants from the East Coast.

At the same time, Crocker decided to take on a new construction foreman, James Harvey Strobridge, a volatile New Englander of Irish descent with a commanding presence, well over 6 feet in height. A thin, hardworking man, he was much resented for his brutal way of keeping order and continually haranguing the men, with or without justification. Even Crocker was sometimes shocked. "I used to quarrel with Strobridge when I first went in," he admitted. "I said, 'Don't talk so to the men. They are human creatures.'" Strobridge would not listen. For him the workers "were about as near brutes as you can get," and he beat them to order with liberal use of his "persuader." Along with his wife and six adopted children, he moved into a passenger car, which would become their home until the race to complete the railroad was over, his figure a familiar and despised presence on the construction site. But when he too failed to hold on to the men, Crocker hit upon what was then regarded as a radical solution: hiring Chinese laborers.

The Chinese had come to California during the Gold Rush, and by 1865, there were 50,000 working on the West Coast. Treated as inferior in every way, the "coolies" were constantly subjected to racist attacks and accused of taking gold rightfully belonging to Americans. "These Chinamen with their shaven crowns and braided pigtails, their flowing sleeves, their discordant language and their utter helplessness, seem to offer entertainment of a nature for which the San Francisco hoodlum is notorious," wrote one observer, the Reverend Gibson. "They follow the Chinamen through the street, howling, screaming and throwing missiles and often these poor heathen reach their quarter of this Christian city covered with bleeding wounds and bruises." Strobridge would have nothing to do with them. "I will not boss Chinese," he raged. "From what I've seen of them, they're not fit laborers anyway. I don't think they could build a railroad!" But Crocker persisted, and in 1865, he hired his first 50 Chinese workers.

Despite Strobridge's predictions, the Chinese labored tirelessly, without a complaint. Slowly, this "degraded race," as Stanford had once described them, earned the respect of the engineers, and Crocker decided to employ them in larger numbers. It began to look as if the Chinese had come to the Sierras to build another Great Wall. "They were a great army laying siege to Nature in her strongest citadel," wrote the *New York Tribune* journalist Albert Richardson, on a visit to the Sierras to see the work. "The rugged mountains looked like stupendous anthills. They swarmed with Celestials, shoveling, wheeling, carting, drilling and blasting rocks and earth while their dull moony eyes stared out from under immense basket hats like umbrellas."

Crocker's army began to take on the western slope of the Sierra Nevada, forging a path through the foothills and constructing giant trestles. Towering sometimes 100 feet above the ravines and up to 500 feet in length, these fragile wooden structures seemed precarious as they closed the distance between the ridges and valleys. With the help of the Chinese, by September 1865, the Central Pacific had reached the 54-mile mark, but it would take another year to cover the next 13 miles to Dutch Flat, over a punishing ascent of 1,200 feet requiring bridges and the circumnavigation of the sheer granite cliff of Cape Horn, which towered almost 2,000 feet above the American River. Beyond this, the incline increased again by 2,485 feet, at which point another gorge, a series of tight curves, and the first of a series of tunnels awaited the Chinese. Crocker sent teams ahead to start work on the tunnels, with men blasting their way from both ends of the Summit Tunnel simultaneously, using up to 300 kegs of gunpowder a day.

With winter approaching and the weather worsening by the day, Crocker and Strobridge decided they had to speed up work on Cape Horn. They hit upon a high-risk strategy. At the suggestion of the Chinese headmen, who had used this method in their homeland for many years, Chinese workers were lowered down in tightly woven baskets onto the cliff face. As they bumped and nudged the rock, mere playthings in the wind, the Chinese struggled to keep their balance and avoid a very quick death either by gunpowder or a long fall thousands of feet below. They patiently bored into the hard rock with hand drills, crammed in the explosive, and, when the fuse was lit, accepted their fate as they were jerked up to safety—or not, as luck would have it. It was extremely risky, not only for those suspended high above in the baskets, decorated with good luck symbols, but for the men clearing the route below, who were constantly exposed to the danger of avalanches crashing down from above. Many of the Chinese workers, who had been imported like livestock from the fields of Canton, lost their lives during the assault on Cape Horn.

By the spring of 1866, the men moved deeper into the Sierras to join the teams who had already begun work on the tunnels. Blasting through the granite was proving slow work, progressing at only 7 to 12 inches a day despite the armies of men involved. With growing impatience, Crocker and Strobridge decided to experiment with a relatively new explosive formula called nitroglycerine. This blasting oil had been patented only a few years earlier in Europe by an engineer called Alfred Nobel, and it was believed to be five times more powerful than powder explosives. This would be the first time it would be used in a project of this scale.

But nitroglycerine also had a reputation for being unreliable. A spectacularly violent explosion had torn apart a ship in Panama that had been transporting the blasting oil, killing 50 men. Then, on April 16, 1866, while in transit in a San Francisco warehouse, a box containing the blasting oil suddenly exploded. Twelve people died instantly and many others were severely injured. Fear of this delicate substance, volatile to the point of being unmanageable, was causing panic. There was a public outcry; it was the "calamitous sensation of the day," declared the *Sacramento Union*. Subsequently, its transportation in this form was banned. Crocker and Strobridge, however, faced with the impenetrable granite of the Sierras, were desperate and not easily dissuaded. They hired their own chemist, James Howden, to mix the formula on site. The blasting oil as mixed by Howden was over 95 percent pure and could not be handled carefully enough. Even a random spark from a drill bit could cause an explosion. At first the results were spectacularly successful, but in 1866, after an accident in Tunnel 8, the Irish workers refused to touch the substance. Only the obedient Chinese were prepared to persevere—until tragedy struck once more. A charge went off unexpectedly and several Chinese were killed. At this point, the use of nitroglycerine was brought to an abrupt halt.

Gunpowder, however, was not as effective as Howden's mixture and progress in the tunnels was reduced to a foot a day. It proved impossible to use with wet rock and tended to shatter the granite into larger pieces, which then had to be broken up further. Within a few months, Crocker and Strobridge decided to experiment with the oily yellow liquid again. The most dangerous work fell to the Chinese, who would lower the substance in bottles, as they had done previously with gunpowder. The work rate doubled, but "many an honest John went to China feet first" observed one engineer grimly. It was only when Strobridge himself lost an eye in a nitroglycerine explosion, when a Chinese shovel hit an unexploded charge, that the Central Pacific had finally had enough. "Bury the stuff," Crocker ordered bitterly.

Quite apart from the explosives, another enemy was waiting in the wings as the worst winter on record descended on Crocker's army during 1866. Driving snow and howling winds whipped through the high passes of the Sierras, reducing visibility to a foot or two. Most Chinese came from the south and had never seen snow. Its dangers took them by surprise, and many died from pneumonia and frostbite. An even greater threat came from the avalanches created by explosives, or from blizzards that blotted out the world for days at a time. These avalanches buried the work along with the men, restoring the mountains' pristine white emptiness. The dead would not be revealed until spring. The men endured 44 storms through that winter, and a total snowfall of over 40 feet.

As a consequence, snow sheds were built to shelter the men and the tracks, enabling the men to work on through the blizzards and storms. As Crocker would later discover, the sheds were extortionate to construct— about $2 million for nearly 38 miles of shedding, a cost Judah had not accounted for because he had underestimated the depth of snowfall. While track laying continued slowly under the snow sheds, the men concentrated on tunneling, which was only possible by burrowing a passage from their sheds to the tunnel entrance, in temperatures of 20 degrees below freezing. Thus, the Chinese lived a strange, twilight existence, buried in a twisting labyrinth of dark tunnels beneath the snow. Apart from a few airshafts, there was little evidence of their existence.

By now, some 14,000 Chinese were working on the mountains and still progress was disappointingly slow. As pressure increased, the white men came to fear Strobridge "as much as they did the Chinese devil." During the late spring of 1867, even the obedient Chinese were finding it increasingly hard to take his orders. They had to endure the utter misery of endless goading and discrimination from bullying overseers throughout relentlessly long working hours with no concern paid for their safety as they dealt with nitroglycerine and gunpowder. Countless numbers were lost in the explosions and the winter blizzards—countless because they were considered so expendable that no records were kept of their deaths. It was later said that at least 10 tons of Chinese bones had been shipped back to their home country for burial. And despite all this, they continued to work until they dropped. Finally, in June 1867, they had had enough: 2,000 workers went on strike. Their demands were modest: their pay increased to 40 dollars a month, the working day reduced to 10 hours, and an end to "the right of overseers of the company to either whip them or restrain them from leaving the road when they desire to seek other employment."

It was a time of crisis for the Big Four. Only 100 miles from Sacramento, and with work on Tunnels 8 and 9 under way, Crocker knew they still had 53 terrible miles to go before finally breaking free of the Sierras, while the Union Pacific seemed to be flying across the prairies. Although some of Crocker's men had traveled ahead to start laying track in the valley, only a continuous line entitled them to government grants. Once again there was a possibility that they would be beaten by financial problems. As for the Chinese, Crocker would not let them hold the company to ransom. There was no alternative but to stop their supplies of food and opium. Isolated in the dangerous peaks of the Sierras near the Donner Pass, the company intended to starve the Chinese into compliance.

BY THE SUMMER of 1867, leaving the lawlessness of Julesberg, Nebraska, behind them, the Union Pacific headed to Wyoming. They had covered almost four times as much ground as the Big Four's team and were determined not to lose such a lead. Their goal was now Salt Lake City, Utah, the only major settlement along the route. Persecuted for their religious beliefs, the Mormons, led by Brigham Young, had made this desolate piece of desert their promised land and had turned Salt Lake City into a thriving metropolis. The closer the Union Pacific team moved to Salt Lake City, the nearer they came to supplies and fresh manpower to enable them to stay ahead of their rivals; but first, they had to cross the most feared Indian territory in the country.

While the Native Americans had initially watched the white men's progress from a distance, it soon became clear to them that the iron road was to be a permanent structure running through *their* land. In theory, treaties protected their land, but the railroad was an exception, one that represented much more than just mere lines of iron to the natives. The white man was eradicating their way of life, desecrating their relationship with the buffalo, dismissing their timeless world of season and ritual. With their very survival under threat, they had to retaliate.

On an August night in 1867, a party of Cheyenne warriors lay in wait to attack the workers four miles from Plum Creek depot, east of North Platte. Unseen, obscured by a ridge of land, they watched as a freight train passed by. "We said to each other, it looked like a white man's pipe when he is smoking," recounted Porcupine, a member of Chief Turkey Leg's Cheyenne, who had never seen a train before. Having just been attacked themselves, and robbed by soldiers at nearby Ash Creek, the Cheyenne were looking for revenge. "We said amongst ourselves, now the white people have taken all we had and made us poor, and we ought to do something," Porcupine continued. They let this first train pass before riding down to investigate the iron rails. Using telegraph wires, they strapped a pile of loose ties to the rails and waited. "If we could throw these wagons off the iron they run on and break them open," said Porcupine, "we should find out what was in them and could take whatever might be useful to us."

The break in the connection in the railroad's telegraph line sounded the alert, and six unsuspecting repairmen, led by a young English signalman, William Thompson, were sent down the line from Plum Creek in a handcar to repair the damage. The Cheyenne lay in wait all around them, hidden in the prairie grass. As soon as the men in the handcar came under attack, they accelerated, hoping to outrun the danger. In the gathering dusk they failed to see the ties on the track in the dusk and struck the blockade at full force. The Cheyenne set upon the railway workers in a frenzied onslaught, killing all but one. William Thompson ran, blinded by terror from the scene, a bullet wound in his arm. An Indian on horseback quickly caught up and clubbed him to the ground with his rifle. Believing he was dead, he proceeded to scalp the injured signalman.

"He took out his knife," Thompson later recalled, "stabbed me in the neck, and making a twirl round his fingers with my hair, he commenced sawing and hacking away at my scalp. I just felt I could have screamed my life out. It felt as if the whole head was being taken right off." Despite the excruciating pain, he knew enough to keep completely still and quiet, pretending to be dead. After what seemed like half an hour, the warrior mounted his horse; and as he turned to leave, Thompson saw his own scalp, his long blond hair streaked with blood, fall off the warrior's belt into the grass, a few feet away.

Lying in the grass, trembling with shock, Thompson could only watch in silent horror as the braves made a mounted attack on the next freight train that came down the line. The driver, Gregory Henshaw, was spared a mauling by the Indians when the train was derailed by the blockade and the cab burst into flames, burning him to death. The engineer was not so lucky, first falling into the hands of the Indians before being thrown back into the wreckage to burn. "A terrible yell announced to the scalped and trembling man [Thompson] the fate of the engineer and the fireman," wrote Henry Morton Stanley in the *Omaha Weekly Herald*. Their charred remains were later taken back to Omaha in "two small boxes, thirty by twelve inches."

It was after the sixth and seventh killings that Thompson was able to make his escape. Taking advantage of the Cheyennes' increasing state of intoxication, courtesy of the whiskey they found in their freight raid, he crawled in the grass to retrieve his scalp and made his escape to Plum Creek, all the while accompanied by the fear that his grisly torturers would discover him. Stanley met Thompson soon after the attack and recorded his full statement. At his side, Stanley observed, in a pail of water, "was his scalp, about nine inches in length and four in width, somewhat resembling a drowned rat." Thompson desperately hoped that surgeons would be able to stitch the scalp back to his head. They failed; but miraculously he survived.

Though the Cheyenne celebrated a small victory that day, they were soon to experience a devastating revenge. Dodge appealed to the government for more help in dealing with the problem. "We've got to clear the damned Indians out," he warned, "or give up building the Union Pacific Railroad. The government may take its choice." Having pledged to protect the railroad workers, General William Tecumseh Sherman, now commander of the Military Division of the West, wrote angrily: "The more we can kill this year, the less will have to be killed the next war, for the more I see of these Indians the more convinced I am that they all have to be killed or be maintained as a species of pauper."

IN THE SIERRAS in late June 1867, the Big Four were still trying to coax their striking laborers back into the tunnels. Eventually, Charles Crocker gave them an ultimatum: they could continue their protest until the beginning of the next working week, but if they failed to return after this, they would be heavily fined for keeping horses and foremen idle. "On Monday morning at 6 o'clock the whole country swarmed with them," he recounted jubilantly, "and we never had so many working before or since as we had on that day." The Chinese had lost their battle and went back into the Summit Tunnel with stoic fatalism.

The years of work and endless insurmountable difficulties had begun to take their toll on Crocker. "I became so that my wife used to be afraid of me," he admitted. Unlike the other members of the Big Four, Crocker was the man on the front line in the Sierras, and the railroad sapped every last ounce of his considerable energy. And so it was, on the morning of August 3, that Crocker became almost light-headed with relief as the two teams working from opposite ends in the Summit Tunnel finally pierced the rock and met in the middle. Despite endless curves and slopes, the route was a mere two inches off center, achieving a degree of accuracy that would prove difficult even with modern technology. It was a great moment for Crocker and his army. The cries of excitement and jubilation echoed along the tunnel. "They have met with some infernal hard rock," Crocker wired Huntington triumphantly, "but you can toot your horn!"

After five years of backbreaking work, the Central Pacific had come only 131 miles to the Union Pacific's 700, but now, with the Sierras behind them, they could finally race downhill and across the flat Nevada plains. "By God, you must work as you have never worked before," Huntington wired his friend. "Our salvation is you—work on as if heaven were before you and hell behind." Hell was indeed behind, and determined to make up ground, Crocker responded to Huntington's words of encouragement by promising to increase the pace to a mile of track a day. It seemed nothing would stop him, not even Indian attacks. When threatened by Shoshones and Piutes that summer as they wound their way down the Truckee Canyon, Crocker struck an imaginative peace treaty with them, promising free rides on the railroad in return for their cooperation. "They have been friendly ever since," reported an impressed Huntington, "and in more than one case they have given the company notice of washouts on the road."

The next challenge for the Big Four was to plan ahead for Salt Lake City. In the spring of 1868, Leland Stanford was dispatched to negotiate a $1 million contract with Brigham Young to provide laborers for grading the track from the Nevada border to Weber Canyon, Utah. Approaching from the other side of the Mormon city, the Union Pacific had the same idea, and made an even better deal with Young, securing 4,000 Mormon workers for grading, tunneling, and construction.

In their urgency to cover ground, both companies began cutting corners wherever possible, throwing together dangerous bridges and sections of track, many of which would have to be rebuilt later. So great was the prize that graders for both companies, working 150 miles ahead of the tracklayers, graded straight past each other along parallel paths for over 200 miles. "Durant was going to the Pacific Ocean, I believe," said Huntington wryly. In fact, both companies were well aware of the "mistake," as both teams of graders were working within sight of each other, but each was reluctant to sacrifice the potential $32,000 per mile. With no meeting point agreed at this stage, there was nothing to stop them. It has been widely reported that violence soon erupted between the rival teams as the Union Pacific workers attacked the Chinese, who fought back using explosives, but these allegations are unconfirmed.

It would be, however, a winter of violence for the Union Pacific tracklayers, many miles behind. They were to face unexpected problems in the last and worst of the Hell on Wheels towns, Bear River City. Here the locals had turned on the gamblers and gunfighters, and the town was governed by a vigilance committee that seemed bent on hanging half the population. Spoiling for a fight one night, 200 drunken railroaders attacked a newspaper office with pickaxes and sledgehammers, only to be ambushed by armed vigilantes as they left. Fifty-three men died that day and many more were injured in what proved to be the state's most violent incident, excluding Indian attack.

Leaving Dodge to pick up the pieces, Durant gave orders to push the men harder, until they were laying an impressive seven and three quarters miles of track a day. Scenting victory, Durant cabled Crocker, betting $10,000 that the Central Pacific could not beat this record. Crocker jumped at the chance. For the Union Pacific 1868 would prove to be the most expensive year, whereas for the Central Pacific it would be the most profitable, with government aid now pouring in. Crocker's mood was good as he accepted Durant's wager with relish. Durant couldn't have guessed he would not have much longer to be making extravagant wagers.

As it happened, a sharp journalist from the *Chicago Tribune*, Charles

Francis Adams, Jr., had been on the trail of Durant's dealings. In January 1869, he published a damning article in the *North American Review* exposing to an astonished public the corrupt practices of the Credit Mobilier and the Union Pacific. He pointed out that the mysterious Credit Mobilier, "is but another name" for the Union Pacific. "The members of it are in Congress; they are trustees for the bondholders; they are directors; they are stockholders; they are contractors; in Washington they vote the subsidies, in New York they receive them, upon the plains they expend them, and in the 'Credit Mobilier' they divide them."

Adams portrayed an image of "ever-shifting, ever ubiquitous characters" involved in a classic case of insider dealing. "They receive money into one hand as a corporation and pay it into the other as a contractor." He revealed that at least $180 million of public money was missing from the railroad's accounts. Although the railroad will undoubtedly be "the most powerful corporation in the world," he concluded, "it will also be the most corrupt." His article caused a great stir in Washington, and journalists hurried to investigate the financial dealings of Durant and his curious company, the Credit Mobilier. What they would find would make the front pages more exciting than news of the railroad.

Congress was losing patience with both companies and their reluctance to name a meeting point. On March 1869, the very day he was sworn in as president, Ulysses S. Grant threatened to withhold funds until a meeting point was agreed. It was decided that Promontory Summit, north of the Salt Lake, would mark the end of the journey. This point, 690 track miles from Sacramento and 1,086 from Omaha, was to become the celebrated site where tracks from east and west would unite to form one continuous line across the continent. The preparations for a grand ceremony began almost at once; and with the momentous date set for May 8, less than a month away, Stanford and Durant set about ordering the most lavish locomotives possible for their triumphant arrivals.

Crocker's mind was on other matters: a certain wager he had made months earlier and fully intended to win. His men would set a record that could not be broken, he declared. The date was set for April 28, he announced, and would be known as the "Ten Mile Day." For the task he selected his best men. In return for their superhuman effort, he promised four days' wages for one day's work—although most would have done it for the challenge alone. Every man knew his place: 850 would lay track, while other teams would unload rails, lay down ties, and operate hand-cars. As dawn broke, a crowd had assembled to watch, including Stanford, Strobridge, and the Casement brothers. A train whistle signaled the start, and within the first eight minutes, "with a noise like a bombardment," the Chinese had unloaded enough rail for the first two miles, and eight powerfully built Irishmen laid the 560-pound rails at an incredible rate, while rail straighteners and spikers slotted in behind. And so they went on—save an hour for lunch—until the whistle blew at dusk. They had laid 10 miles and 56 feet in 12 hours. Crocker had won the wager of his life, and in the process the Central Pacific had taken the rail within five miles of Promontory Point.

The Central Pacific team laid their last rail bar on May 1, 1869, with the Union Pacific hard on their heels. All that remained was to spike the last rail and the great Transatlantic Railroad would be complete. In a curious twist, the Union Pacific, after leading all the way, was upstaged at the eleventh hour as the Big Four found themselves waiting for their rivals—and for Durant in particular. In addition to the indignity of being forced to abandon his impressive locomotive when heavy rains brought down a shoddily built bridge at Devil's Gate, there was further scandal when he was kidnapped by his own workers. At the lonely station of Piedmont, his train was ambushed by 300 furious tie-cutters who had not been paid for several months. Unless he wired for money, they threatened—and the kidnappers were demanding at least $200,000—Durant would be taken into the mountains as their hostage.

With rumors rife about exactly what had happened to Durant, the great Promontory Summit ceremony was delayed by two days. Unable to wait, on May 8, Crocker went ahead with a party back in Sacramento. "My heart is full—boiling over," he declared to an adoring crowd. At 11:00 the town exploded into a deafening cacophony of sound when more than 20 engines, including the Governor Stanford, let rip with their whistles for a full 15 minutes; then fire bells and steam whistles from foundries, steamers, and machine shops for miles around blew and rang out for a further hour. At 3:00, after a grand procession and a splendid lunch, the speeches began, which were full of praise for the townspeople of Sacramento, "who loaned us their money then gave us their confidence, which was better than money." Most grateful of all was the long-suffering Mrs. Crocker. "Charlie," she had once threatened, "if you were going to build another railroad, I should just want to get away from you."

As for Durant, disaster was temporarily averted when an initial payment of $50,000 was wired ahead and he was released. But where Durant was concerned, there were always rumors, and when his train finally arrived at Promontory Point on May 10, it was said that he had staged his own kidnapping in order to claim a kickback of 10 percent of the ransom money! The rumors, however, did not appear to bother him as he graciously managed a dignified entrance, stepping down from his train, wearing a black velvet coat and colorful necktie, his appearance a lesson to all in the art of confidence. Crowds had been gathering since early in the morning on that bright spring day as construction trains packed to overflowing with workers whistled their way into Promontory, everyone in a celebratory mood as they waited to see the two lines joined.

The high point of the ceremony was the historic driving of the last spike. First came the driving in of the ceremonial spikes, including one made of pure gold and inscribed: "May God continue the unity of our country as this Railroad unites the two great Oceans of the world."

These were duly removed and kept for posterity in velvet-lined boxes. Finally, the last real spike could be hit home. After a slight altercation over who—Leland Stanford or Thomas Durant—would drive in this final spike, in the end Stanford stepped forward. His hammer and the spike were wrapped in cable wire so that the whole nation might share in the moment, cannons and guns waiting to fire in celebration. It was agreed that Durant would drive in a second spike seconds after Stanford. But according to folklore it is said that both missed, and it was left to Dodge, Strobridge, and Jack Casement to drive home the spike.

At last the news was wired around the nation, and a telegram was sent to President Grant: "THE LAST RAIL IS LAID! THE LAST SPIKE IS DRIVEN! THE PACIFIC RAILROAD IS COMPLETED!"

It was an emotional day; even Strobridge, in a magnanimous gesture, showed his appreciation of the workers he had once scorned. "When the other guests arose from the table," reported the *Chicago Tribune*, "Mr. Strobridge introduced his Chinese foreman and laborers, who had been with him so long, and took the head of the table, making some excellent remarks, and inviting them to the banquet." The crowd responded with yet more ecstatic cheers. Only one person was missing who would have most wished to see this ceremony: Theodore Judah, the man whose single-minded vision had made the railroad a possibility. "The spirit of my brave husband descended upon me," wrote his wife, Anna, that day—which was also her wedding anniversary—"and together we were there unseen."

Six years after its inauguration in Sacramento, Sherman's "work of giants" was complete, and with it the economy, politics, and landscape of the country were changed forever. The Big Four rose above their early struggles to become delightfully rich. Dodge continued to work for the Union Pacific, his reputation and bank balance unscathed. The scheming Durant, as happens in all good stories, schemed until the end—and beyond. Financial ruin and social disgrace would follow him beyond the grave, with a $20 million claim on his estate. As for William Thompson, after trying and failing to have his scalp sewn back on, he donated it to the Union Pacific Railroad Museum, where it remains to this day.

Just as President Lincoln had envisaged, the railroad helped to create one nation, undivided. The centuries-old worn buffalo trails began to disappear as the country shifted gear to embrace the iron and steel of the future. And as the great unstoppable iron road took the long journey east from Sacramento through the towering peaks of the Sierra Nevada, passing the clear waters of lonely Lake Donner, then on to the towns with names so evocative of American history—Green River, Medicine Bow, Plum Creek—and across the Great Plains of Nebraska and on to the fertile Missouri, it created one country more surely, more simply, and quickly than treaties, proclamations, and wars ever did.

CHAPTER 6

THE PANAMA CANAL

They never achieve anything who do not believe in success. . . .

VICOMTE FERDINAND DE LESSEPS

A RIBBON OF WATER, gold in the western sun with the dark shapes of large ships gliding quietly from the Atlantic to the Pacific, was the dream. Ever since the young Spanish explorer, Vasco Núñez de Balboa, went down in history as the first European to see the Pacific Ocean in the early sixteenth century, the inviting prospect of uniting these two great oceans seemed within man's grasp. That enticingly thin strip of land, the Isthmus of Panama, separating North and South America, approached from the Atlantic through a ring of tropical islands in the Caribbean Sea, was the place where the beguiling challenge of uniting east and west, dawn and dusk, looked so easy to achieve.

There were plenty of warnings, however, that such a dream was well beyond the means of man. As early as 1625, a Jesuit scholar, Josephus Acosta, argued that a canal was not only dangerous but would invite the wrath of God:

I believe there is no humaine power able to beate and breake down those strong and impenetrable Mountains, which God hath placed betwixt the two Seas, and hath made them most hard Rockes, to withstand the furie of two Seas. And although it were

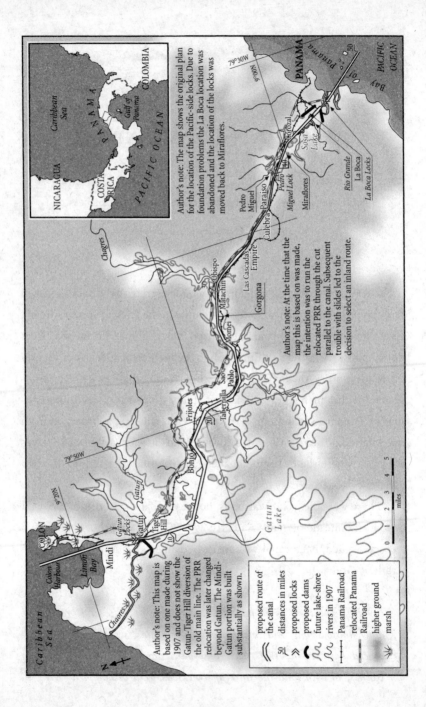

Author's note: This map is based on one made during 1907 and does not show the Gatun–Tiger Hill diversion of the old main line. The PRR relocation was later changed beyond Gatun. The Mindi–Gatun portion was built substantially as shown.

Author's note: At the time that the map this is based on was made, the intention was to run the relocated PRR through the cut parallel to the canal. Subsequent trouble with slides led to the decision to select an inland route.

Author's note: The map shows the original plan for the location of the Pacific-side locks. Due to foundation problems the La Boca location was abandoned and the location of the locks was moved back to Miraflores.

proposed route of the canal
distances in miles
proposed locks
proposed dams
future lake-shore rivers in 1907
Panama Railroad
relocated Panama Railroad
higher ground
marsh

Caribbean Sea

Caïon Harbour
Limon Bay
Chagres
COLÓN
Mindi
Gatun Locks
Gatun
Tiger Hill
Bohio
Tabernilla
Sao Pablo
Frijoles
Momei
Matachin
Gorgona
Las Cascadas
Empire
Obispo
Culebra
Pedro Miguel
Paraiso
Miraflores
Pedro Miguel Lock
Corozal
Sosa Lake
Rio Grande
La Boca
La Boca Locks
PANAMA
Bay of Panama
PACIFIC OCEAN
Gatun Lake
Chagres

79°50W
79°30W
9°20N
9°00N

NICARAGUA
COSTA RICA
PANAMA
Caribbean Sea
Gulf of Panama
COLOMBIA
PACIFIC OCEAN

0 1 2 3 4 5 miles

possible to men, yet in my opinion they should fear punishment from heaven in seeking to correct the workes, which the Creator by his great providence hath ordained and disposed in the framing of this universall world.

But by the late nineteenth century, such doubts had long been dispelled. The Suez Canal was completed in 1869, and now a canal across the isthmus seemed not only possible, but essential, due to the boom in world trade. The canal would shrink the whole world, as ships would be able to sail through Suez, proceed across the Atlantic, and then pass into the Pacific through the new canal, which would cut straight through Central America at its narrowest point. The rewards for such an enterprise were boundless; thousands of sea miles would be saved and ships would no longer have to struggle against the wild seas around Cape Horn. The long and dangerous journey from New York on the East Coast to San Francisco on the West would be reduced from almost 14,000 miles to just 6,000—saving three weeks' sailing time. A mere 50 miles cut through the mists of green exotic flora would bring a fortune and glittering reputation to the person who met this challenge.

Such a man was ready and willing to cut the continent in half and make a road of water, safe for shipping. He was a Frenchman, Vicomte Ferdinand de Lesseps, the hero of his fellow countrymen for turning the speculative plan of the Suez Canal into a practical reality. Not a young man anymore, but still charming to all who met him, he seemed to embody hope, the real possibility of success. His optimism carried the day at the Société de Géographie in Paris in May 1879, at a meeting of the world's most expert engineers.

In this age of scientific achievement, de Lesseps told the delegates, the great undertaking was within reach at last: a second massive canal to complement the work he had already completed in Egypt with the building of the Suez Canal 10 years earlier. Many nations had been interested over the years in building such a canal, but the technical difficulties involved had always outweighed the will to do it. Now, fired by the triumph of Suez, de Lesseps was confident of success. The seductive chimera beckoned and France would win the glory.

Ferdinand de Lesseps was at the height of his powers in 1879 when he became leader of the Panama enterprise. A spectacularly successful entrepreneur and diplomat, he had brought unexpected affluence to many of his countrymen when he had masterminded the building of the Suez Canal, and was now affectionately known by his supporters as *Le Grand Français*, a man, it was said, "who walked almost as an equal among kings." At 74, he had the vigor and energy of a far younger man, and was married for the second time to a woman half his age, whose beauty and elegant sense of fashion only served to redouble interest in de Lesseps himself as they held court together in his Parisian mansion. That he had powers of persuasion beyond the ordinary was well known; many who came within the enchanted field of his magnetic power seemed lost to reason, able only to see the great—and achievable—glorious destiny.

Later it would be said that he had "too much confidence in his luck, in his superiority, in his star," but at Suez, de Lesseps had gained an unshakable belief in his own powers. He had defied all his critics who predicted that the 105-mile passage through the desert was too daunting a task to complete; he had even raised the capital for the project. When the Rothschild banking house had demanded a high commission to finance Suez, he had simply gone directly to the French public to gather the necessary capital by issuing shares. The Suez Canal had been financed to the tune of 200 million francs by 25,000 small investors—and when the share price had soared, their fortunes were assured. On his triumphant return from building the Suez Canal, he had had honors heaped upon him at home and abroad. Thus, any new project from de Lesseps was bound to arouse interest.

His son, Charles, always a steadying force, but a man cut from the same cloth of charisma and charm, begged his father to have nothing to do with the venture in Panama. "What do you expect to gain at Panama—money? You will concern yourself no more with it at Panama than you did at Suez. Glory? You have acquired so much that you can afford to leave some for others. . . . Certainly the Panama project is a magnificent one . . . but what risks those who direct it will have to run! You succeeded at Suez by a miracle; should one not be satisfied with one miracle in the course of a single life, instead of counting upon a second?" But his father had already made up his mind. The grandiose vision, uniting two oceans with a silver ribbon of water, was all-consuming, impossible to resist; and with Charles at his side, no one could dissuade him.

De Lesseps felt that the experience he had gained in Egypt would be invaluable for the new enterprise. He aimed to build a sea-level canal, exactly as he had at Suez, in effect, cutting a level path right across Panama, from the Atlantic to the Pacific Oceans—"from sea to shining sea." The best route, he thought, was to follow the Panama railroad, which had been built across the isthmus during the Gold Rush of the 1850s. Furthermore, his proposed canal would be only half the length of Suez—a mere 50 miles.

The route across the isthmus was, however, a wickedly deceptive obstacle course. The suffocating jungle filled every moist inch with massive green trees, smothering creepers, and gaudy exotic flowers. Looping through it all and meandering endlessly back on itself was the Chágres, a slow and threatening river just waiting for the rainy season when it would be transformed into a raging torrent, capable of dislodging bridges and creating havoc with recurring floods. There were black, bottomless swamps lying in the path of the canal, too; and worst of all, the Cordilleras mountain range made an implacable barrier.

Speaking at the 1879 conference at the Société de Géographie, a pessimistic note was struck by the engineer Godin de Lepinay, who, unlike de Lesseps, had experience of working in Central America. De Lepinay warned that it would be impossible to excavate the necessary amount of earth and rock, and that if they attempted to cut through the Chágres River while in flood it would certainly inundate the canal. Instead, de Lepinay suggested damming the Chágres to create an artificial lake and building a system of locks to raise ships up from sea level. But these proposals proved much more expensive to implement than de Lesseps's sea-level canal, and in the end *Le Grand Français* won the day. De Lesseps had absolute faith that science would work miracles. "Science has declared that the canal is possible and I am the servant of science," he reasoned. His next task was to raise the necessary funds—not something to trouble him unduly. Money would surely fall like manna from heaven when he asked for it, as it invariably did.

In July 1879, de Lesseps went on a lecture tour of France to raise private subscriptions; he also appealed directly to investors in Britain and America. But matters went less smoothly than he had hoped. Following the Franco-Prussian War, there was a mood of economic caution amongst potential investors. De Lesseps also encountered criticism from the press and opposition from powerful French bankers who resented his methods of raising finance. As a result, the subscription was a disappointment, raising only a fraction of the 400 million francs he required. De Lesseps would have to launch a second campaign to secure the money he needed.

Undeterred by the failure of the public subscription, he decided to visit Panama and assess the canal site firsthand. He set out in December 1879 with a large retinue that included a technical commission of French engineers and members of his family. His entourage was courteously received and much feted; de Lesseps himself was given an ecstatic welcome. Soldiers lined the route and French flags were in abundance as his party drove to the finest hotel. A dinner given in his honor turned into an all-night party, which overflowed into the square bejeweled with hundreds of colored lanterns; the music and dancing continued until dawn.

But distracted by all the excitement of ceremonial banquets, fireworks, canoe rides, and expeditions, in their short visit, de Lesseps's party failed to gain detailed knowledge of the isthmus and its conditions. The technical commission, however, did endorse de Lesseps's preference for the position of the cut through the Cordilleras and the sea-level method.

An incident occurred when digging the inaugural spade of earth on New Year's Day 1880 that underlined the proceedings with an ominous and somber note. A position had been chosen for the Pacific end of the canal, in Panama Bay not far from the mouth of the Rio Grande; but as champagne flowed freely, the departure of the boat to take them to the site was delayed. When they finally arrived, the site was underwater of the rising tide. De Lesseps refused to be defeated by the sea. As an alternative, a box containing sand was brought on deck, along with his 7-year-old daughter, Ferdinande, who dug a token spadeful. The innocent simplicity of the child's action made a striking contrast to the grandness of the venture in which man dared to achieve mastery over the forces of nature.

With no funds, however, work could not begin on the canal. Over the coming months, de Lesseps continued waging a one-man propaganda war in France. He launched a private newspaper dedicated to the Panama project, undertook extensive lecture tours, and even paid substantial bribes to politicians and influential newspapers. Finally the press voiced their support, with *La Liberté* positively scolding its readers: "The Panama canal has no more opponents. . . . Oh, ye of little faith! Hear the words of Monsieur de Lesseps, and believe!" *Le Gaulois* went even further: "Success is certain," its readers were assured. "One can see it, one can feel it. The enthusiasm is universal." At last, in December 1880, the company's stock was successfully floated as more than 100,000 people clamored for shares. With the support of small investors, de Lesseps now had the capital to proceed.

The Compagnie Universel du Canal Interoceanique was formed to create the canal, with Ferdinand de Lesseps and two of his sons, Charles and Victor, on the board of directors. During 1881, de Lesseps painted a rosy picture of the future to the board, announcing with absolute certainty that the canal would open within seven years. The first French engineers soon began to arrive in Panama and were greeted by an environment very different from home. The tropical jungle was intensely beautiful, populated by strange and exotic flora and fauna. De Lesseps considered it the most enchanting region in the world: "A perfect paradise; there is no botanical garden in the world which could surpass its beauty." He had seen the isthmus at its best, however, during the dry season. Panama was a far more formidable opponent than it had appeared on de Lesseps's first visit. The French engineers, although dedicated and highly qualified, had to contend with a number of problems apparently unforeseen by de Lesseps and his technical commission.

Foremost amongst these was the climate. Not only was it hot and humid, but the rainy season, which lasted eight months of the year, brought torrential downpours for which the Europeans were completely unprepared. The dense jungle was alive with venomous snakes, big cats, and myriad insects. They also had to contend with filthy living conditions and inadequate health services. The key cities, Colón, on the Atlantic coast, and Panama, 50 miles southeast on the other side of the isthmus looking out on the Pacific, were dirty, dilapidated, and rampant with disease and poverty. Despite this, in the spring of 1881, the French team began by surveying and clearing a path for the canal, several hundred feet wide across Panama from Colón to Panama City, and building accommodations and a new hospital for the workers. Employees began to stream steadily into Panama from France, eager to share in the great honor of building the canal. Within a few weeks, however, they encountered yet another terrifying obstacle: yellow fever.

Despite Ferdinand de Lesseps's claim in 1880 that "Panama is an exceedingly healthy country," virtually every tropical disease lay in wait: typhoid, cholera, smallpox, malaria, and dysentery all were constant threats. But it was yellow fever that caused the most panic, perhaps because it seemed to "favor" those new to Panama, sometimes within days of their arrival. The symptoms began with fever and uncontrollable shivering, and then moved on to severe back pain, which many described as like being on the rack. A terrible thirst, the mouth on fire, was the next symptom; then the skin turned yellow, until the victim, coughing up black blood, waited, with a dropping body temperature, for the inevitability of death. As many as 70 percent of those infected succumbed. Word of the dreaded "yellow jack" spread rapidly back to France, especially after white-collar employees began to contract the sickness. De Lesseps, who was based in France, seemed adept at not hearing anything that was inconvenient. Eternally optimistic, he assumed that all difficulties would be overcome.

Many believed that yellow fever only afflicted those who led an immoral life. Drinking, gambling, and visiting houses of ill repute—all readily available diversions in Panama—were thought to make a man more susceptible to the fever. This view was so firmly held by the canal company's new director general, Jules Dingler, that he thought it quite safe to bring his whole family—including his daughter's fiancé—together with all their thoroughbred horses, to Panama. At first all went well, with picnics and riding for the family, but then his beautiful 18-year-old daughter, the center of the family, came down with yellow fever and died, while they watched in disbelieving horror. Madame Dingler, in uncontrollable distress, wanted to leave immediately and take her son with her, but Dingler insisted that their staying was a matter of honor. They could not break the trust that de Lesseps had put in them. Some weeks later, their 21-year-old son, too, succumbed to the disease and died; his death was followed by the daughter's fiancé. Madame Dingler, like a woman lost, nevertheless continued to accompany her husband on his trips.

Despite the string of tragedies that struck his family, Dingler threw himself into "*la grande enterprise*." Noted for his organizational abilities, he was tough and did not hesitate to ruthlessly cull those he considered shirkers. He had planned three main teams of workers, one dredging at each end of the canal, in Panama Bay on the Pacific side, and Limón Bay, near Colón, on the Atlantic. A third team began in the Cordilleras with an all-out assault on the lowest point of the mountains at the Culebra Cut. It was soon apparent that the planned excavation of 74 million cubic meters was a substantial underestimate. A sea-level canal, 72 feet wide and almost 30 feet deep cut in a swath across Panama, would need 120 million cubic meters of rock and earth to be removed. De Lesseps accepted Dingler's figures, yet made no change to his forecast of a completion date of 1888.

Meanwhile, news of the rising death rate did not seem to deter more French employees from arriving on a weekly basis. By May 1884, the workforce stood at over 19,000. One of those to make the journey across the Atlantic that year was the 26-year-old French engineer Philippe Bunau-Varilla. For four years, since hearing de Lesseps's lecture in Paris, it had been his great ambition to work on the canal. On the voyage over, a chance meeting with Jules Dingler, who was returning from a short period of leave, set him on his future course. Impressed with the young man's enthusiasm and technical knowledge, Dingler soon appointed Bunau-Varilla in charge of the vast excavation at Culebra and the Pacific area of the canal.

Culebra was the lowest point of the Cordillera mountain range that completely blocked the path of the sea-level canal. Cutting through the mountain at this point was proving an impossible task. The cut was to be nine miles long and several hundred feet deep; but with inefficient equipment and no thought given as to where to place the spoils, the work had, after several years, made little impact. Dredges, steam shovels, and locomotives were brought in from Europe and the United States; but at Culebra, much work also had to be done by hand, which made progress there slower than elsewhere along the line. Bunau-Varilla, however, seems to have approached Culebra as a battle to be won whatever the cost.

Quite apart from the difficulties they faced at Culebra, it soon became clear that there was no easy solution to the problems posed by the Chágres River, a willful, volatile river flooding frequently in the eight-month rainy season. The route of the canal was to cross the twisting Chágres possibly several times, but no real solution had been reached as to how to contain this mighty river when in flood. Bunau-Varilla and his colleagues soon discovered that torrential rains, falling from the mountain ranges in the wet season, could cause the Chágres to rise more than 35 feet. It was apparent that such a violent flow of water rushing into the finished canal would pose a massive threat to both ships and human life. Some believed that building a dam across the Chágres at Gamboa could solve the problem, but no detailed investigations had been carried out. De Lesseps, himself, had only seen the river during the dry season when it was just few feet deep. When questions were asked about the canal, he continued to state confidently that science would triumph over nature. "We have in Panama five divisions of engineers, and every one is convinced that we cannot fail to attain the desired end," he declared in 1885.

In fact, the problems at Culebra were beginning to prove insurmountable, spreading doom amongst the men as they came to realize the aptly-named Culebra—meaning snake—would never loosen its stranglehold on them. The very twisting nature of the cut, the degrees and angles of fall, the idiosyncrasy of the rock with its layers of clays, shales, and gravels, would always conquer their most innovative and determined efforts. Eventually, even de Lesseps could not close his eyes to Culebra, the snake. In the rainy season, huge mudslides were caused when the upper layer of clay became saturated and slithered downwards into the cut, crushing or covering anything in its path. As work progressed, the slides grew worse, dumping hundreds of tons of sticky wet mud back into the canal. The spoil banks, too, proved increasingly unstable, causing mini-avalanches that often stopped work completely by burying the tracks of the dirt trains.

Nevertheless, more willing recruits continued to pour into Panama. They came on boats from France, filled with patriotic fervor; from elsewhere in Europe, with their technical expertise; and from the surrounding regions of Jamaica, Colombia, Venezuela, Cuba, and America countless Black laborers flooded in. By the summer of 1884, however, the death rate had climbed to over 200 workers a month. The streets saw a constant stream of funeral processions, and trains ran daily to the cemetery at Monkey Hill. Many of the poor never went to the hospital for treatment; they simply died where they fell, to be covered over with a load of spoil. Bunau-Varilla, now in command of a much greater proportion of the canal, told of ships standing empty in the harbor at Colón because all the crew had died. As the crisis deepened, no one could deny the danger. Jules Dingler was once again overtaken by grief as his wife, who had stayed loyally by his side, succumbed to yellow fever over Christmas 1884 and died an agonizing death. After the funeral, completely overwhelmed with sorrow, now having lost all his family, he took all his beautiful horses up into the hills and shot them, one by one.

By summer 1885, the excavation was falling seriously behind schedule, yet back in France, the ever-confident de Lesseps insisted that the problems were surmountable and that the work was already half done. Certainly, most small shareholders still seemed to be behind him, but Panama shares had begun to fall a little on the Bourse—the Parisian stock exchange—and the canal company had become vulnerable to attack from rogue traders. There was also criticism in the newspapers, with the *Economiste Français* warning against "the most terrible financial disaster of the nineteenth century," and the financial presses in New York and London loud in their condemnation. De Lesseps's solution was to pay another visit to Panama to raise morale there and to increase confidence in the Panama project at home in France.

It was clear the workers on site idolized him, as he was received with enthusiasm; but not everyone shared de Lesseps's optimism. Among the laborers for a short time was the painter Paul Gauguin, who had fallen on hard times in France. He was appalled by the working conditions in Panama, and found the physical labor on the canal exhausting and the heat stifling. "I have to dig from five-thirty in the morning to six in the evening, under tropical sun and rain. At night I am devoured by mosquitoes," he wrote plaintively to his wife. It was so far from being the noble project that he had envisaged that after two months of backbreaking work in the unremitting heat, he left to find a better life.

Progress was constantly impeded by torrential rain, mudslides, floods—even the wildlife was hostile. On one occasion in December 1885, Philippe Bunau-Varilla was surveying from a canoe the aftermath of another devastating Chágres flood when he saw the strangest phenomenon. His eye was held by a curious change in the green color of the tops of the tropical trees and foliage: "A zone of about a yard above the water formed a striking contrast. It was *black* [emphasis mine]." As they moved closer, he realized in horror that "the leaves and branches were totally concealed by enormous spiders of the Tarantula species . . . there were millions and millions of spiders chased by the inundation." Just as he was struggling to steer a path away from the tarantulas without capsizing, "wild cries" came from another passenger in the boat. A two-foot coral snake had slithered into the canoe and "its head was in my left hand coat pocket." Torn between his violent repulsion of the deadly reptile and the need to avoid any sudden movement that might capsize them, he slowly managed to coil the snake around a folded umbrella and lower it into the water.

Faced with continual setbacks, by 1886, barely a quarter of what Dingler had hoped to achieve had been accomplished. Even de Lesseps's supporters were beginning to consider that a completion in the late 1880s was beginning to look like the wildest optimism. Although de Lesseps worked tirelessly to raise more money to advance the canal, rumblings continued in both the French and the American press. A cartoon that appeared in *Harper's Weekly* asked: "Is Monsieur de Lesseps a Canal Digger, or a Grave Digger?"

Bunau-Varilla was well aware that the project was in trouble. Echoing the earlier proposals made by Godin de Lepinay, he had already come up with a solution. He tried to persuade de Lesseps that rather than a sea-level canal, a lock-and-lake canal that raised ships up above sea level would be the easiest and quickest way to finish since less excavation would be required. A clever feature of the plan was that, according to Bunau-Varilla, such a canal could then be converted back into a sea-level waterway, using a new technique he had devised of underwater blasting and dredging. "I was holding in reserve the provisional lock canal," Bunau-Varilla later admitted, "to be transformed later, by dredging, into a sea-level canal, in case the impossibility of digging the sea-level canal directly in the dry should be demonstrated." But it would be a long time before de Lesseps would accept even this compromise, and by then it would simply be too late.

The financial problems continued—three bond issues had failed to raise the money needed to persist with the canal. In seven years, the French had succeeded only in excavating a great ditch, 70 feet wide, from Colón on the Atlantic side for 11 miles inland. Along the rest of their canal route were clear signs of their advance, but the dreaded impasse at Culebra Cut had been reduced only by around 150 feet. On top of this, the death toll was estimated at over 20,000. With the adverse publicity and shortage of funds, in December 1888, de Lesseps placed his hopes in a last, desperate issue, but once more Panama stock was falling on the Bourse, and he failed to win support.

In February 1889, the canal company went into liquidation, wiping out the hard-earned savings of 800,000 private investors who had placed their implicit trust in de Lesseps. Under duress, de Lesseps was obliged to telegraph the isthmus to call a halt to the work. So far, over $280 million had been spent, and the scandal that followed stunned the French nation and led to the downfall of its government. It was the largest financial collapse of the nineteenth century. Bunau-Varilla was utterly disgusted by what he saw as a loss of faith: "Just when the canal enterprise deserved for the first time to be regarded as an absolute certainty, public confidence was withdrawn from it," he despaired.

Shortly after the collapse of the company, Ferdinand de Lesseps's health began to fail as well. He lived in quiet seclusion, fiercely protected by his wife and family, and was steadily becoming more frail and confused. But it was not to last. The day came in 1891, when accused of fraud and breach of trust, both Ferdinand de Lesseps and his son Charles were summoned to appear before the court to answer charges. Charles was fearful for his father, who did not seem to grasp the gravity of the situation; but, somehow, at the appointed hour, *Le Grand Français* rediscovered some of his former inner strength and magical powers of communication and left for the private hearing, his face alive with all the old charm and vivacity. "But the reaction was not long delayed and it was terrible," continued Charles. "From the next day for three weeks my father never spoke and never left his bed." He seemed tormented by his train of thought and became weaker by the day.

As the weeks passed into months, his confusion increased and he seemed to be largely unaware of what was going on around him. This was perhaps just as well, since Charles was eventually arrested and thrown into prison to await trial. The bewildered old man greatly missed his son, with whom he had always had the closest relationship, and accused him of neglect. In prison and unable to help his father, Charles learned from de Lesseps's doctor that his father's "wish to embrace me had become such a fixed idea that it could bring on fever, which could prove fatal." Eventually Charles was given compassionate leave, under police escort, to visit his father, and a short-lived scene from the family past was revisited as they all had lunch together in the formal dining room, with Ferdinand de Lesseps taking his place at the head of the table, benign and reassured. He did not seem to be aware of the policemen; and when his son left, he had no idea that it was to return to prison.

By the time the case came to court in 1893, Ferdinand de Lesseps was too unwell to attend, whereas Charles had to endure two lengthy trials before the verdicts were finally announced. The de Lessepses' lawyer, Henry Barboux, pointed out that there was no evidence that either Ferdinand or Charles had tried to gain financial advantage from the canal project, pleading that neither father nor son could ever be dishonest and that their only sin was overoptimism and misjudgment. His words fell on deaf ears. While heavy sentences were imposed on several directors of the canal company, the two de Lesseps fared the worst. Both were sentenced to five years in prison and ordered to pay a heavy fine. Ferdinand was sentenced in absentia, much to his son's distress. And although on appeal the sentences were later reversed for technical reasons, Charles was unable to pay the fines arising from the second trial and fled into exile for several years.

Ferdinand de Lesseps lived on quietly in his country house, in a gentler world; he had no memory of his disgrace and the prison sentences. Attended by his wife, Louise-Hélène, who still loved him deeply, he had no suspicion of the terrible difficulties their son had suffered. Charles was back with him when he died one winter afternoon, aged 89, when the grand old man, whose charm had seemed able to conquer all obstacles, simply fell asleep for the last time.

To his wife, Louise-Hélène, he was a hero to the last, a perfect husband and Frenchman without blemish. He had been motivated only by honor, and what crime was that? she asked. "It would be impossible to find another man who, having accomplished what he did at Suez ... would have retired from it with empty hands, with no provision for the future of his large family—a sacrifice in which I, as well as my children take pride." While she revered this as a mark of his illustrious character, the de Lesseps family now had so little money that even Ferdinand's funeral expenses had to be paid by the Suez shareholders.

Although there can be little doubt that the Panama Canal would never have come into existence without the efforts of Ferdinand de Lesseps and his engineers, for many years afterwards, in France, the word "Panama" was synonymous with corruption and disgrace.

FERDINAND DE LESSEPS was dead but there was still one man who believed in him, whatever his reputation, and would not let his dream die. Philippe Bunau-Varilla was totally committed to the Panama Canal and deeply disappointed that the French had lost confidence in the project. So, in late 1900, he sailed for the United States to try to persuade the American government to take on the canal.

Following the assassination of President William McKinley in 1901, America had a new president, Theodore Roosevelt. Roosevelt was very keen to make the United States a world power and regarded a canal through Central America as essential to secure naval supremacy. He was only too aware of the time it had taken one battleship, the *Oregon*, to reach the Atlantic around the Horn during the Spanish-American War— a journey of more than two months, which saw her arrive barely in time to be of service.

Many of the American senators, however, held the view that the canal should be routed through Nicaragua, rather than Panama, sensing perhaps that the ghost of French failure haunted the Panama route. Bunau-Varilla was determined to persuade them that it made far more sense to go through Panama: the route was shorter, straighter, would require fewer locks, and was already supported by the Panama Railroad. He made no mention of the difficulties that had led to the failure of the French attempt, and instead set out to convince the senators of the drawbacks to the Nicaraguan route, including the risk of volcanic activity. In June 1902, shortly before the decisive vote, Bunau-Varilla sent each senator a Nicaraguan stamp featuring a famous volcano of the region, shown in an alarming state of full eruption. A few days later, the Senate voted in favor of a canal through Panama, and the Americans bought the French company for $40 million.

Unfortunately, the Colombians, who governed the territory crossed by the canal, were reluctant to ratify the treaty that gave the Americans the right to build this canal in Panama. Roosevelt was furious that they should try to hold him to ransom, and decided to support Panama's independence movement, sending warships to both Colón and Panama City to block the sea approaches. Troops were also sent in to protect the railroad.

Bunau-Varilla realized that, as a Frenchman, he was free to play a role that an American diplomat could not, and it seems likely that Roosevelt was well aware what he would report back to the revolutionaries. And so it was that at the Waldorf-Astoria Hotel in New York City that modest and deferential Frenchman, who had an inner core of steel where French honor was concerned, arranged a secret meeting with Amador Guerrero, the spokesman for the Panamanian rebels. Bunau-Varilla offered Guerrero support on behalf of the U.S. government and money to pay his troops. He also suggested that on completion of the coup he himself should be invited to act as Panamanian minister in Washington. When American warships appeared in Limón Bay to protect their interests, the revolution was as good as complete.

Bunau-Varilla, as "Envoy Extraordinaire and Minister Plenipotentiary," was now in a position to act for the new Panama Republic. On its behalf, and with remarkable speed, he negotiated a treaty with Roosevelt's Secretary of State, John Hay, which came to be known as the Hay-Bunau-Varilla Treaty. There is still a feeling in Panama to the present day that Bunau-Varilla sold out the rebels to the United States, but the deed was done. Three days after the treaty was signed in February 1904, Bunau-Varilla resigned his post and returned to France. "I had . . . assured the resurrection of the great French undertaking," he said with pride. "France banished the child she had brought forth, but I had succeeded in having it adopted by a friendly nation."

THE FIRST YEAR of American administration of the canal was, however, even less successful than that of the French. In March 1904, Roosevelt appointed the members of the Isthmian Canal Commission to superintend the work. The position of chief engineer was filled by a Chicago railroad engineer, John Findley Wallace, who had many years of experience in the American Midwest. Unfortunately, Roosevelt had not foreseen the frustrations that would arise as decisions were made by an absent commission whose members all rarely agreed. The sheer quantity of paperwork caused massive delays in obtaining the simplest items of equipment, and when Wallace cabled Washington to protest, he was sharply reprimanded by one of the commission that sending cables cost money.

Wallace found the task facing him in Panama overwhelming. Everywhere, the ghostly signs of the French failure were in evidence. More than 70 dredging machines were found, covered by jungle where they had ground to a halt five years previously. A village was also discovered, buried in the green half-light like a sunken ship, with machinery and various supplies all subsumed beneath abundant foliage. Spurred on by Roosevelt's insistence that they must "make the dirt fly," Wallace plunged into digging at Culebra even though he had no clear plan, often using the plant and equipment left behind by the French. But he proved to be an uninspiring leader, remote from the men and haunted by fear of disease; moreover, the commission held such tight reins on the funds that he had no easy means of improving conditions for the workers.

The Americans were already aware of the dangers of disease, as they'd suffered in a terrible military experience in Cuba in 1898, when 80 percent of Americans occupying the island contracted yellow fever, and countless others died of malaria and typhoid. "I feel that the sanitary and hygienic problems . . . on the Isthmus are those which are literally of the first importance," Roosevelt warned in 1904, "coming even before the engineering." It was essential to find the right medical man to take charge of the sanitation at Panama. The leaders of the commission soon alighted on a most fortunate choice: Colonel William Crawford Gorgas, an army doctor and leading expert in tropical diseases.

Everything about Gorgas was kind and disarming, as though something of the gentleness of childhood had never left him; silver-haired quite early, his face benign. Yet for all his self-deprecating charm and easygoing temperament, he was rigorously disciplined. His father had been a general in the Civil War, and as a child he had hoped to be a soldier, too. When this proved impossible, he found that he could still be in the army if he trained as a doctor. He had met his wife, Marie Doughty, while they were recuperating from severe bouts of yellow fever, which left them both with an immunity to the disease. Though it was not Gorgas himself who discovered how yellow fever was transmitted, he was credited with eliminating the disease from Cuba by thorough, painstaking work. He had little doubt that he could perform the same role in Panama, given adequate power and support.

Under the French regime, a desperate battle had been waged against malaria and yellow fever. Impressive hospitals had been built and excellent medical staff brought in, but the simple fact was that, at the time, knowledge of tropical diseases was not sufficiently advanced to win the battle against them. It was only now, at the end of the century, that a doctor working in India, Ronald Ross, found evidence that the *Anopheles* mosquito spread malaria; while in Cuba, Dr. Walter Reed identified *Stegomyia fasciata* as the carrier of yellow fever. Gorgas was also aware that it was the female insect that transmitted the diseases from infected individuals to healthy ones via its bite, and that mosquitoes could breed only with access to standing water. Armed with this knowledge, he had a chance of controlling the deadly diseases.

Unfortunately, the theory of mosquito-borne infection had not yet found widespread acceptance, and Gorgas's masters on the Canal Commission were deaf to his appeals for adequate supplies and staff to carry out his campaign. The belief in "miasma," fumes arising from the swamps and landscape, as the source of sickness was still in vogue, as was the belief that moral and clean-living individuals would be immune to disease. Thus, when Gorgas arrived at Colón in June 1904, it was with only limited supplies of essential items and a small team of workers. "When they landed at Panama to engage in the mighty task of ridding the jungle of disease," Marie Gorgas said later, "they had little more than their own hands and their own determined spirit to work with."

There was plenty to keep them occupied. Mosquitoes were abundant in almost all buildings, breeding in the rainwater barrels and the drinking-water jars used in most households. Gorgas set up his headquarters at Ancon hospital, only to discover that there were no wire screens on the windows and that patients already infected with yellow fever or malaria were not shielded by mosquito nets to prevent the spread of disease. He also found that plants in the attractive hospital gardens were protected from umbrella ants by stone rings filled with water, and that the bed-legs in the wards themselves were also frequently resting in water to ward off ants—all providing perfect breeding places for mosquitoes. So numerous were the insects that the doctors and nurses had to struggle to protect themselves as they worked, and most contracted tropical diseases themselves at some point. "Probably if the French had been trying to propagate yellow fever," said Gorgas, "they could not have provided conditions better adapted for this purpose."

Within a few months of his arrival in Panama, Gorgas made a trip to Washington, D.C., eager to win support to deal with the crisis, only to be told that his ideas on the mosquito were "balderdash." "On the mosquito you are simply wild. . . . Get the idea out of your head," said Governor George Davis, another member of the Canal Commission. Gorgas, however, would not be browbeaten into taking what he believed was the wrong course and returned to Panama even more determined to continue with his plans. "That persistence which had always been his chief asset . . . forced him to the task," observed his wife. Nevertheless, he was all too aware that time was running out; large numbers of workers due to arrive in Panama from the States and the West Indies would soon provide fresh blood for the voracious insects.

As December 1904 approached, as many as 3,500 men were laboring in the eternal mud, with little comfort provided when the day's work was over. Accommodations were scarce and, as a consequence, expensive, with some men forced to resort to living in shantytown huts. Moreover, the food was of doubtful quality, of little variety, and often overpriced, while most water was undrinkable. Less than a year after the arrival of the Americans in Panama, reports of the dreaded "yellow jack" were reaching the ears of the press in the United States and causing panic amongst the workers. "The rank and file of the men have begun to believe that they are doomed, just like the French before them," commented Gorgas in frustration.

As the hospitals began to fill with the sick and dying, many workers decided to go home. The sight of funerals passing through the streets every day only added to their sense of hopelessness. In June 1905, Wallace returned, disillusioned, to the United States to tender his resignation, opting instead for a lucrative job in a healthier climate. Roosevelt soon appointed in his place a new chief engineer: John Stevens.

Stevens was a brilliant choice. Growing up on a small farm, he had soon learned to be self-sufficient, and in the absence of a formal education, had taught himself chemistry, physics, and math. He had a survival instinct, too, gained on trips to the Rockies, where he had faced blizzards, attacks by wolves, and cold nights at below 40 degrees, while single-handedly finding a pass that saved 100 miles for the railroad. Tough, enigmatic, and strikingly good-looking, his reputation enhanced by the myths that surrounded his previous exploits, he was, at 52, the preeminent railroad man in America, and had now been given carte blanche by Roosevelt to "get the job done."

Stevens arrived in Panama on July 26, 1905, to a situation worse than he had been led to believe. The whole project was suffused with an air of pessimism. Most workers had fled from the dreaded yellow fever; those remaining were demoralized. The lack of any organization, coordinated effort, or even a coherent plan astonished him. He made it his first task to get on site and discover for himself the fundamental flaws that were plaguing the epic venture. He was soon seen everywhere, puffing on his cigar, inspecting, reassuring, listening to everyone, gathering information, while quietly coming to terms with the massive problems left behind by Wallace.

He was horrified by the inadequate plant and equipment being used, much of it inherited from the French. "This is no reflection on the French," remarked Stevens, "but I cannot conceive how they did the work they did with the plant they had." The outdated machinery was not up to the massive demand made of it, yet Wallace had been quite happy to use it. Stevens decided that it was pointless to proceed with anything less than the most up-to-date machinery available. The railway track being used to carry spoil away from the excavations would also have to be much heavier. In addition, he was conscious that the men were demoralized and "scared out of their boots." Declaring that "the digging is the least thing of all," Stevens's first act was to order all work on Culebra Cut to be stopped until proper preparations had been completed—though he knew that this approach would not find favor with the press corps at home, who expected, at last, "to see the dirt fly."

Stevens also decided to throw his weight behind Gorgas and the sanitary campaign—though he could not quite bring himself to believe in Gorgas's newfangled ideas about mosquitoes. "Like probably many others I had gained some little idea of the mosquito theory," he commented, "but, like most laymen, I had little faith in its effectiveness, nor even dreamed of its tremendous importance." Nonetheless, he knew the eradication of disease was absolutely vital to the completion of the canal, and he was impressed by Gorgas's conviction that he could achieve results. When some members of the Canal Commission back in Washington started a campaign to have Gorgas removed, Stevens demanded that he stay and saw to it that Gorgas was generously equipped with everything and everybody he needed to launch an all-out attack on yellow fever. Gorgas later acknowledged the vital importance of Stevens's backing: "The moral effect of so high an official taking such a stand at this period . . . was very great, and it is hard to estimate how much sanitation on the isthmus owes to this gentleman for its subsequent success."

So began the most expensive and intensive sanitary campaign ever fought. Gorgas set in motion a systematic program of fumigation, with orders to place tons of insecticide powder and fumigation materials each month, in addition to implementing copious cleaning agents and equipment. A team of several hundred men set to work to fumigate all the private houses in Panama, a process sometimes repeated several times despite the skepticism of the inhabitants. Huge quantities of wire mesh for door and window screening also arrived to stop mosquitoes spreading the disease. Above all, Gorgas concentrated on preventing the female mosquitoes from depositing their eggs. Teams of men sprayed tanks, drains, cesspools—wherever water accumulated—with a film of oil to seal the surface; they even went into the surrounding countryside to fill in ditches and pools of standing water or to spray them in oil. Finally, the cities and settlements were provided with running water so that no one would need to keep water in open containers. Meanwhile, better quarters were being built for the workers.

One of Stevens's most important contributions to the project was his recognition that the railroad was vital to the construction of the canal. It had been put forward by Bunau-Varilla and others as one of the reasons for the superiority of the Panama route, but Stevens recognized that insufficient use was being made of it; indeed, the railroad had fallen into complete disrepair. In the course of his first year, Stevens developed the railway extensively. Lightweight tracks were replaced with much heavier rails, and two tracks were laid so that there could be flow of traffic in both directions simultaneously. Engine sheds and repair shops were built, and a new workforce was recruited and devoted entirely to the running of the railroad. Since Stevens also ensured that supplies of fresh food were brought in via the improved railroad, there was now the superstructure to provide better food, which in turn helped to raise morale.

Stevens then turned his attention to the dirt trains taking the spoil away from the excavation site. He recognized that the operation at Culebra Cut could be transformed by using the railroad much more systematically. "Neither the inspiration of genius nor our optimism will build this canal," he declared. "Nothing but dogged determination and steady, persistent, intelligent work will ever accomplish the result . . . We are facing a proposition greater than was ever undertaken in the engineering history of the world." For Stevens, the steady, persistent work began with an attempt to keep the dirt trains constantly rolling at the cut day and night. The spoil was to be carried immediately away from the excavations, both speeding up the work and reducing the risk of mudslides.

Gorgas, meanwhile, was at last beginning to see his labors bear fruit. Just a year and a half after his arrival at Panama, the doctor was able to tell a group of his staff, as they gathered in the dissecting room at Ancon hospital, to "take a good look" at this corpse, as the unfortunate victim was likely to be the last case of yellow fever they would see. Yellow fever was, at last, under control; and by the end of the year, it had been eradicated. Gorgas was aware, though, that his work was far from done. It would be some time before the incidence of malaria could be dramatically reduced because the *Anopheles* mosquito, which was less of a domestically bred insect than *Stegomyia*, required a more far-reaching extermination campaign.

Stevens's improvements to living conditions had made a great deal of difference to the welfare of workers and their families. By 1906, around 24,000 men were laboring at the canal. With the improvements to the railroad, and bigger and better machinery in hand, Stevens was now ready to resume excavation in the Culebra Cut. Soon, work there was moving faster than ever before, yet he was still unsure of the final plan. Agreement had not been reached on whether or not he was building a sea-level canal. The time was fast approaching when a decision had to be made, and Stevens, having thoroughly familiarized himself with the task at hand, was firmly of the opinion that the only way to tackle Culebra was via a lock-and-lake canal—the same conclusion reached by Philippe Bunau-Varilla, before him.

Now Stevens found himself summoned to Washington to explain his views to a Senate committee. He made a convincing case there, explaining that even with their best efforts, it would take an unacceptably long time to cut through Culebra down to sea level, and that the costs would be unjustifiable. Despite the progress he had already made in the removal of spoil, he thought it likely that extra excavation required for a sea-level canal would increase the risks of further mudslides. Most critical of all, Stevens had now seen the Chagres in full flood. He knew that a lock canal would allow the damming of the river, giving control over any sudden rise in water level, rather than risking it flooding into the canal. He declared a sea-level approach to be "an entirely untenable proposition."

His plan, therefore, was not to hack through Culebra, but to go over it, damming the Chágres at Gatún to create the largest man-made lake in the world, 85 feet above sea level. Approaching traffic would be lifted up to the level of the lake via a system of locks, then lowered down on the other side through more locks to return to sea level. There was a good deal of opposition to Stevens's plans. Doubt was expressed about the safety of the earth dam he proposed, which, it was thought, could collapse under the weight of water, as the Johnstown Dam had done in 1889 when a complete town had been washed away, along with 2,000 lives. But Stevens's case was compelling; he brought all the years of his engineering wisdom into the reasoning, and slowly, he brought America's president around. Roosevelt was impressed and gave Stevens an overwhelming vote of confidence.

The president was so full of enthusiasm, so eager to show how much he valued the work, that he decided to visit Panama for himself. The idea of an American president leaving his home country to look at a canal caused something of a sensation, and his trip aroused widespread interest. To enable Roosevelt to see the immense problems caused by the rainy season, and to fully appreciate the challenges facing Stevens and his men, it was decided that he would travel in November.

Roosevelt seemed to enjoy the trip, demanding to be shown absolutely everything, including the mess halls, where he insisted on eating with his wife on one occasion when he was expected at a grand ceremonial lunch. He won the hearts of the laborers by inspecting the workings for himself, even posing in the pouring rain at the helm of a massive Bucyrus steam shovel, clad in an immaculate white suit with a huge smile on his face.

Roosevelt was delighted with everything he saw. From his point of view, the venture was clearly an undertaking worthy of America, and his speech reflected his pride: "You here who are doing your work well in bringing to completion this great enterprise, are standing exactly as a soldier of the few great wars of the world's history. This is one of the great works of the world. It is a greater work than you yourselves at the moment realize." Even the workers' living quarters and the hospitals did not escape the president's attention; and he was vocal in his praise for William Gorgas and his battle against disease. By the time Roosevelt left, everyone—including Stevens—was exhausted by the pace and thoroughness of his visit.

Confidence in the project had never been so high; and confidence in Stevens in particular, still gilded by the glory of the recent visit from Roosevelt, was unlimited. It was a shock, then, that without warning, just three months later, Stevens resigned.

"When I accepted the offer to become chief engineer of the canal, I told him [the president] that I could make no promise to remain until the work was fully completed," Stevens pointed out later, "but that I would stay until I had made its success certain, or had proved it to be a failure. . . . I knew that the work was so well organized that competent engineers could be found to carry it on to a successful completion."

To many people, this was not a full explanation of his decision, so there was much speculation that Stevens had broken down under his responsibilities, that he was unhappy about his relationship with Washington, or that there was some hidden scandal. Roosevelt was amazed and angry. Enigmatic to the last, Stevens left, saying he would never reveal the true reason for his departure.

IN THE WAKE of Stevens's resignation, Roosevelt decided to place the work "in charge of men who will stay on the job until I get tired of having them there, or till I say they may abandon it. I shall turn it over to the

Army." He chose Lieutenant Colonel George Goethals, a military engineer whose knowledge of hydraulics and lock design and construction surpassed that of Stevens—though Stevens's experience of railroad engineering and his grasp of the need to improve living conditions for the workers had laid the foundation on which Goethals could build.

At last Roosevelt had found the right man for the enormous task ahead. Goethals, with his serious approach to life, seemed older than his 48 years. From an impoverished immigrant Flemish family, he had made his way by hard work alone, and expected complete dedication to the task at hand from those around him. He was tall and fit, with very blue eyes that seemed to look straight into the heart of any problem with an uncompromising directness, as though his mind were always at attention. He had no interest in the fame that came with his position and was concerned only with his goal, setting his workers a fine example of devotion to duty. He was also very conscious that, as Roosevelt had commented, there could be no question of him, an army man, changing his mind or quitting his post.

But Stevens proved to be a difficult man to follow, and Goethals found that his predecessor had been extremely popular with the men. These men now dreaded the army-style regime that they feared would be imposed by the new colonel and his military engineers. On his arrival in Panama in March 1907, Goethals endured a number of snubs. To counteract, he assured his listeners at a welcoming dinner that he was impressed by the work Stevens had done and that, although the army was now in charge, he had no intention of bringing in a more dictatorial approach: "I now consider that I am commanding the Army of Panama," he declared, "and that the enemy we are going to combat is the Culebra Cut and the locks and dams at both ends of the Canal, and any man here on the work who does his duty will never have any cause to complain of militarism."

The scale of the task before him was not lost on Goethals. "The magnitude of the work grows and grows on me . . . but Mr. Stevens has perfected such an organization so far as the railroad part . . . is concerned, that there is nothing left for us to do but to just have the organization continue in the good work it has done . . . ," he commented with characteristic magnanimity. He knew, however, that several years of hard work still lay ahead. Fortunately, Goethals, who had made his reputation in the Corps of Engineers, had already built both locks and coastal defenses and was uniquely qualified for the job. He immediately changed the organizational structure, dividing the canal into three zones, rather like military divisions: the Atlantic Division, including Gatun locks and dam, under Major Sibert; the Pacific Division, including the Pacific locks, under Sydney Williamson; and the Central Division, including Culebra Cut, under Major Gaillard. He hoped that competition would arise between the army men and the civilians on the workforce, thinking this could only be good for the canal.

For his part, Dr. Gorgas soon found his relationship with Goethals rather different from that which he had enjoyed with John Stevens. His wife, Marie Gorgas, had no time for the man. "His passion for dominating everything and everybody he carried to extreme lengths," she declared. "He was utterly lacking in adaptability." The incident that provoked her fury concerned the trimming of grass and vegetation, which Goethals felt was being done at unnecessary expense by Gorgas's department. Goethals wanted to place this job under the control of his own quartermaster; and although Gorgas guarded his territory fiercely, Goethals was eventually to have his way. But Gorgas would have the last laugh. "Do you know, Gorgas, that every mosquito you kill costs the United States Government ten dollars?" demanded Goethals, exasperated. "But just think," replied the doctor, "one of those ten-dollar mosquitoes might bite you, and what a loss that would be to the country!"

Amongst the workers, however, Goethals soon began to earn respect. He gave them access to him to air their grievances every Sunday morning at his office—no mean achievement considering that by 1907 there were 32,000 men on the payroll, and by 1910 nearly 40,000. The employees found him fair, and were impressed by his willingness to devote time to their individual problems and requests. For Goethals, the other reward for his hands-on approach was a detailed knowledge of the practical problems encountered by his labor force.

Life for the canal employees was now quite different from their experience under the French. The American workers and their families were mostly in white-collar jobs—teachers, nurses, doctors, engineers, administrators—and their pay and accommodations were, in many cases, better than they could have enjoyed back home. There was no longer a feeling that they lived in the outback. Married quarters were luxurious; servants were cheap; and amusements were abundant, including glamorous parties, dances, and department stores (which stocked everything imaginable, from clothes pegs and mops to high fashion, and all at discount prices). On Sundays in the dry season, families took full advantage of the climate and beauty of Panama, making expeditions to beaches and places of interest.

The West Indian workers did not, however, enjoy the same advantages as the Americans. Though they were by far the majority of the labor force and filled essential jobs such as cook, porter, sanitary operative, and gravedigger, in addition to many thousands laboring on the canal itself, there were sharply defined race rules, which excluded them from many of the privileges available to white workers. The discrimination even extended to the way the different classes of employee were paid—the white in gold, the black in silver. Pay for the "silver" employees was far lower and living conditions considerably less luxurious.

Meanwhile, there were plenty of technical challenges to be overcome. Goethals had been a firm supporter of Stevens's proposal for a lock canal, but there were still many back in Washington who believed that the vast earth dam at Gatun would be unsafe and that the plan to

flood a huge area of the surrounding countryside was completely im-
practical. Nevertheless, Goethals and his men simply carried on with it,
dumping spoil from the excavation to form the basis of the dam, exactly
as Stevens had envisaged. Gradually, it crept across the valley for over a
mile and half until it became so wide, relative to its height, that any damaging land slip was almost impossible. The huge embankment had two
thick outer walls of the dry rock waste from Culebra; the inner core con-
sisted of blue clay that dried as hard as concrete. When, in 1908, a 200-
foot section slipped and sank almost 20 feet, it caused a storm of criticism
in the newspapers, but Goethals was more than able to withstand it.
Insisting that the situation was not serious, he simply had the damage
repaired and carried on.

As the excavation forged ahead at Culebra, work continued day and
night, with dynamite crews playing a vital role, in addition to those man-
ning the rock drills and steam shovels. Over the entire period of the
canal's construction, more than 61 million pounds of dynamite were
used. Between explosions and machinery, the din at the cut was deafening—but it failed to deter the parties of sightseers and tourists who visited regularly and were even offered guided tours. "He who did not see
the Culebra Cut during the mighty work of excavation missed one of the
great spectacles of the age," wrote one awed visitor.

The unprecedented amount of dynamite now being used at Culebra
day and night made it sound like a war zone. Adding to the noise was the
endless roar of heavy machinery moving the waste and the trains continually carting it away. The work continued for seven years. Much of the
success of this effort was due, of course, to bigger and better machinery,
such as the 95-ton Bucyrus steam shovels; but continued efficiency in
transporting spoil, too, made a huge difference, with the construction of
rail tracks being regularly shifted and relocated to keep work moving.

Unfortunately, as the speed of work increased, more and more men
were being killed or maimed in gruesome accidents with dynamite. To

stanch the trend, Goethals issued strict regulations about how explosives should be handled, believing that part of the problem lay with carelessness on the part of the workers. But the dynamite also became volatile from exposure to the Panamanian climate, and accidents were also caused by such things as excavating machinery hitting unexploded charges, or lightning striking. In December 1908, one devastating explosion left over 60 injured and 23 dead. The workers came to call the cut "Hell's Gorge" because of the intense heat and the injuries caused there. Fortunately, though, as the dynamite crews gained experience, accidents became less common.

One difficulty that was never fully resolved, however, was the matter of mudslides at Culebra. They began to be a problem again from October 1907, when a slide at Cucaracha continued for 10 days, slowly, irrevocably, bringing down a weight of earth each day. It was as though half the mountain were on the move, burying everything that lay in its way, destroying tracks, dirt trains, and excavation machinery, and wiping out months of work. Structural, or "deformation," slides added to the problems as the cut grew deeper. Altogether, slides in the cut were to add more than 25 million cubic yards to the total excavation and increased the total time taken by at least a year. To some, the cut appeared to be a task that would never be finished. "No-one could say when the sun went down at night what the condition of the Cut would be when the sun arose the next morning," wrote one of Goethals's advisers, Joseph Bucklin Bishop. "The work of months and years might be blotted out by an avalanche of earth." Goethals, refusing to be deflected from his task, simply issued instructions to dig out each slide all over again.

Arguably the greatest challenge facing the canal builders was the design and construction of the great locks. Goethals had revised the plans for both the width of the canal and the size of the lock basins, which were now to be 1,000 feet long by 110 feet wide—big enough to accommodate a ship the size of the *Titanic*, with room to spare. These locks were bigger than anything attempted before. Another unique feature of the lock basins was that they were to be constructed from concrete, still a relatively new building material and certainly not one that had been used on a project of this size before. Not only was a high degree of accuracy called for in mixing the component parts, but the liquid concrete also had to be delivered speedily and efficiently to the site, where it would be poured into huge molds, or "forms," each around 36 feet. Within these, additional forms were required to shape the necessary passageways, tunnels, and culverts needed for the operation of the locks.

By using a system of cranes, towers, cableways, and huge buckets, Goethals and his men were able to achieve a fast overhead delivery system, which left a free working area on the floor of the lock chambers. More than 2 million cubic yards of concrete were used at Gatun alone. The huge locks presented a strange sight with the exacting precision of their tall straight walls and giant gates, every angle, every line in perfect geometry, and all surrounded by the unruly elements of nature, where no straight line existed.

The moving parts of the locks also had to be specially designed: the great steel double gates and their mechanisms. Varying from 47 to 82 feet in height, by 65 feet wide and 7 feet thick, they weighed several hundred tons each. Nothing like this had been built on such a scale before. Once the locks were filled with water, the gates would weigh comparatively little, since they were hollow in construction and would float, but they would be tested in a dry lock to ensure that the opening mechanisms were up to the job. The designer of these mechanisms, Edward Schildhauer, was working without an established precedent for precision engineering of this size, and commented that the available information was "at variance." When tests were performed, the steel struts and wheels that opened and closed the vast gates operated flawlessly.

The power that drove the mechanisms was electricity—a great innovation at a time when steam- and horse-powered systems were the norm. Some 1,500 electric motors were required to run the canal, so a hydro-electric power plant was constructed near to the Gatún Dam, where the falling water at the spillway would generate enough electrical current to supply all the canal's energy needs. The control system for the locks was worked from a point in the central wall of the upper lock, allowing a single operative to run every operation in the passage of a ship.

By 1913, Goethals found himself in a position where the work was going well and morale was high as the workers saw the end in sight. The locks had taken four years to build, but were finished almost a year earlier than estimated. The damming of the river at Gatún had also been successfully completed, making the mighty Chágres safe forever. The man-made lake would eventually cover about 164 miles of jungle; and as the water slowly rose, submerging many villages and all but the tallest trees, the fresh water lake, a brilliant ocean green in color, framed by the distant blue of land and mountain, looked as though it had always been part of the landscape. Although the mudslides were still a problem, Goethals now elected to finish the excavation using dredges once the cut was filled with water.

On September 26, 1913, the first trial lockage was successfully completed at Gatún, by a small tug decorated with flags and bunting for the occasion. A festive air arose as thousands watched the first vessel make the watery ascent. Goethals was present to see the massive mechanism run like clockwork, taking 1 hour, 51 minutes.

Only a few days later, on September 30, the canal had its ultimate test when an earthquake erupted nearby. It lasted the better part of an hour, with repeated tremors damaging buildings over a large area. A church collapsed and buildings were ravaged. Measuring 6 on the Richter scale, it raised fear about the stability of the newly finished canal. Would this mark its ending all too soon? But the canal proved resistant to any hazard that nature could produce. With great delight, Goethals was able to inform officials in Washington that the canal had suffered no damage at all.

On October 10, America's new president, Woodrow Wilson, pressed a button in Washington and relayed by telegraph via New York and Galveston to Panama a signal to blow up the center of the temporary Gamboa dike and join the Culebra Cut to Gatún Lake. With Culebra now fully flooded and dredges working day and night, the canal was almost ready to be opened to traffic.

The canal proved to be a technical triumph, a true wonder of the modern world. The ultimate success of the project was due to the dedication of a few remarkable men, some of whom were themselves destroyed in the process. No one knew this better than Philippe Bunau-Varilla. He was generous in his praise for Goethals and his achievements in bringing the enormous and unwieldy project to a brilliant conclusion. He was generous, too, in his praise for the United States, saying, "It is a day of glory for the American nation, which generously gives this waterway to the world without any special privilege for herself or her citizens and without any mercenary considerations in view." But his own national pride compelled him to persuade the citizens of Panama to erect a statue to de Lesseps, where, today, it permanently surveys *"la plus belle région du monde."*

Goethals, meanwhile, was quite sure to whom the Panama Canal owed its existence. "The real builder of the Canal," he said, "was Theodore Roosevelt." Roosevelt had set it all in motion and inspired the massive effort. His was the glory, just "as if he had personally lifted every shovel full of earth in its construction." The last of the great engineers to view the canal was John Stevens, who came back unobtrusively one day to marvel at Goethals's magnificent accomplishment, to see the huge ships go by, and the enormous locks working so efficiently, the silent clockwork of their huge doors closing with tight precision.

It had been a herculean task, a world event, far greater than de Lesseps could ever have imagined as he sat in his drawing room in 1879 making plans. It had taken 35 years and cost $639 million dollars. Many had given so much for this strip of water that changed the world; but, sadly, this grand achievement was eclipsed by events taking place that would change the world forever. Europe was waiting in somber mood for the signal of war. It was declared on August 3, 1914, as the canal, a side event now in the shadow of such news, opened for business.

Still, the inauguration was an emotional moment for both Charles de Lesseps and Philippe Bunau-Varilla who, bristling with pride at the part the French had played in it, stood on the deck of the first boat to use the canal, a modest working cement craft called the *Cristobal*. There were no dignitaries, no special pageant, no world leaders, no big speeches. The president declined to come, and the solemn moment passed. And so it was that the oceans of the east and west were mingled at last as the *Cristobal* slid through the quiet waters of the Panama Canal. Twenty-five thousand lives had been stolen by the extravagant dream—the longest fifty miles in history.

CHAPTER 7

THE HOOVER DAM

Now this is just a dam, but it's a damn big dam.

WILLIAM WATTIS, *President, Six Companies, 1931*

FOR CENTURIES THE Colorado River has cut its way for 1,400 miles through barren and difficult land, creating a maze of deep canyons and powerful waterfalls. Its source is in the Rockies, where a million hidden streams cut their way through snowy wastes to mingle with the long and rushing Green River, whose union with the Colorado creates an awesome torrent. On south through the arid plateaus of Utah and the desolate wilds of Arizona, it meets the Little Colorado, carrying its burden from the White Mountains. Now with a fury of water collected from 1,000 square miles, the river twists west to race in a series of rapids through the Grand Canyon and the lonely Black Canyon on the Arizona-Nevada border, whose sides rise sheer to trap the river when in spate. Eventually, across the burning Mojave and Sonoran deserts reaching the Gulf of California and the sea this mighty river carries water from regions of six states, and drains the western slopes of the Rockies. Over 740 billion cubic feet of water pour down the Colorado each year.

During the late nineteenth century, engineers and surveyors began to look for ways to harness the natural power of the Colorado and make this arid region habitable. In 1901, a private company, run by engineer Charles Rockwood, cut a canal and drainage channels into the western bank of the river just north of the point where the Colorado flows into Mexico. Water from these channels was used to irrigate the Imperial Valley in California. With advertisements promising the richest farmland in the States, thousands of settlers flocked to the valley.

For a few years, the scheme worked well, producing magnificent crops and lush grass for fattening cattle; but there was a price to pay for this Garden of Eden in the desert. The canal, blocked with sediment, failed; and when a new canal was dug on flatter land with no head gate to hold the river back when in flood, the annual snowmelt of 1905 turned the river into a roaring tumult, breaking the earth walls of the canal, deluging the land, and submerging all the prosperous farms. An inland sea was created, 30 miles long and 10 miles wide. The river was out of control, and it would take two years to put it back on its original course.

It was all too clear: the unruly river would have to be tamed to control flooding, irrigate the barren lands of southwest America, and provide power. Arthur Powell Davis, head of the U.S. Bureau of Reclamation, the agency responsible for solving the irrigation problems of the West, began to investigate. For years his team studied the area, mapping streams and tributaries, measuring waterflow, surveying the parched deserts of the region. "The river was the enemy," observed one official. "It was something that could kill you or ruin your farm."

Davis, however, whose uncle had been the first person to navigate the entire river, knew its potential, and he had a vision. In his mind's eye he saw a whole string of man-made lakes and irrigation systems stretching like glittering beads on the necklace of the dry canyon country, distributing water where needed. In 1922, he produced a report setting out his bold plan to build a massive dam on the Colorado. A bounty of water and electricity was there for the taking.

The favored position of the dam was narrowed down to two possible sites: Boulder Canyon and Black Canyon. Both appeared to fulfill the criteria needed for a successful reservoir: they offered deep, narrow, sheer-sided walls of hard rock. Finding the very best location was critical. If they built on or near a fault line, it could give way, causing the dam to fail and a mountain of water to sweep over all the towns that lay in its way. The river might find a new path and wander over a floodplain, for years causing havoc, as it had in the Imperial Valley. After exhaustive tests, geologists uncovered faults in the rock of Boulder Canyon that cast doubts on its sustainability. Black Canyon, 30 miles southeast of Las Vegas, Nevada, had superior rock quality, and the narrowness of the gorge with its precipitous cliffs plunging to the river below would be an asset when building the concrete dam.

Although Black Canyon was the final choice, illogically, the project continued to be called Boulder Dam. And it was to be a huge dam, a colossus: the grandest undertaking in which the U.S. government had ever been involved. The engineers in the Bureau of Reclamation produced a design for a concrete arched gravity dam that would rise from the riverbed to a sheer 727 feet, more than twice as tall as the Statue of Liberty, its width at the base an impressive 660 feet of solid concrete. The curving form ensured that the force of the immense weight of water pressing against the dam would be transferred to the massive canyon walls abutting either side. Twin towers at either side of the dam would extract water and deliver it to the powerhouses situated farther downstream at the foot of the dam, where electricity would be generated. Huge spillways 650 feet long and 170 feet deep would ensure that in times of flood, water could not spill over the top of the dam. It was an immense and magnificent scheme—one that also promised employment for many. It would need an army of laborers to turn these lonely canyons into the proposed modern marvel.

After years of lobbying, on June 25, 1929, the Boulder Canyon Project Act was approved by President Herbert Hoover. A few months later, in October, the stock market crashed and the Great Depression that followed lay on the land like a terrible blight. All over America there were lines for jobs; unemployment was running at 25 percent. Hearing rumors that work might begin on the dam in spite of the economy, as the national crisis deepened in the early 1930s, people began to flock in desperation to the region. They came on foot or in their Model T Fords and from the dirt farms where they loaded up their dilapidated trucks with their families and a few essential possessions. The then sleepy little town of Las Vegas, with its population of 5,000 in the middle of the Nevada desert, was overwhelmed. A soup kitchen was opened there for the destitute, but most went hungry. Shantytowns sprang up all around the town. Tents were put up anywhere there was a space; some camped near the graveyard, and hundreds slept "rough" near the station.

The actual site of the proposed dam was 30 miles south of Las Vegas, across the glaring white desert, where the shimmering heat gave endless movement to the distant black hills and the dazzling sand. Anxious not to miss recruitment, many chose to make that forbidding trip on the dirt road directly to the site of the dam, setting up their makeshift dwellings by the river in temperatures of 120 degrees. In the unremitting sun, with the dense black volcanic rock behind them retaining the heat, and not a tree for shade, the sad little houses made of cardboard, empty petrol cans, and old bits of cloth sprang up overnight. Soon there were 1,500 people living in what came to be called "Ragtown."

There were no services of any kind; even the drinking water that came from the river required a long wait for the silt to settle. "The river was too thick to drink and too thin to plough," said the locals. Endlessly tormented by sandstorms, always in fear of snakes and scorpions, the women tried desperately to create a semblance of normality, but the combination of disease and flies and heatstroke took their toll as the people waited and waited for work to begin. "An intangible sense of disappointment vexed me; a sort of feeling of depression and danger," recorded one resident of Ragtown. "Those strange mountains, that oily river, those masses of black rock, those deep crevasses, what could mere man hope to accomplish against that overwhelming fortress built by nature—what dangers, what tragedies lay ahead of their pygmy efforts at mastery?"

Still the flow of workers continued to arrive, people so desperate for work that they just walked out of their houses with anything they could carry. "People came with their kids, they came with everything on their backs. And their cars broke down before they got here, and they walked. No one helped them. I was the only person they could come to," said Murl Emery, who ran a small café in Ragtown. "The government would have nothing to do with them. . . . So we began to feed these people. Some of them had no money. Everything was on honor. It worked. I wound up with only two checks I never collected. I'm satisfied they were two men lost who were not accounted for."

In January 1931, the government advertised for construction firms to bid to build the dam. But it was clear it was beyond the hope of any one firm to take on such an ambitious project. After much discussion, feeling that united they stood a chance, six large construction companies in the West joined forces to offer an estimate to build the dam. They called themselves simply: Six Companies. "Now this is just a dam, but it's a damn big dam," declared their president, William Wattis, in his under-stated style. On March 4, 1931, their bid, at $48,890,955 proved to be the lowest. Six Companies also boasted another winning card: their chief engineer, Frank Crowe.

"I was wild to build this dam," Crowe recalled later. "I had spent my life in the river bottoms." He had followed the tangled fortunes of the enterprise since 1919 and it had become an obsession for him. Building this dam would "mean a wonderful climax, the biggest dam ever built by anyone anywhere." Engineering was in his blood. His wife used to tell a story of how, on their honeymoon in New York, the one thing that held her husband's interest was a new variety of dump truck he discovered on the street. Since 1905, when he had graduated from the University of Maine, Crowe had worked on civil engineering projects all over America.

He was much respected, both by his employers and his workers, some of whom were so loyal that they followed him from venture to venture. Apart from his legendary managerial skills and organizational flair, his innovative ability had him designing time- and labor-saving methods on almost every project. He had designed an overhead cable system for transporting both materials and men quickly on site, which was especially useful when dam building. Not a man to set much store by a grand office and a big desk, he hated paperwork, his own personal maxim being "never my belly to a desk." He was always somewhere on site, anticipating problems, his very presence seeming to confirm the growing myth that he was some kind of superman, ever driven by the tantalizing idea of turning so many tons of crude materials into a massive edifice of complex usefulness and beauty.

One week after Six Companies made its successful bid, on March 11, 1931, Crowe was on site and ready to start. Everything about the dam was a challenge. It would be the biggest dam in the world, holding back with one wall an immense body of water in a lake 150 miles long and up to 580 feet deep. Skeptics doubted it could ever work. The weight of water in the lake would surely invoke an earthquake that would destroy the dam and deluge the land with waves hundreds of feet high. Some even speculated that the sheer weight of the water could throw the Earth out of orbit. Quite apart from whether it would ever work, there was also the question of how it could be built in the first place. Black Canyon was surrounded by mile upon mile of arid canyons and to the south and east stretched burning deserts. There was no proper transport or services of any kind, and yet all the immense paraphernalia of men and machines somehow had to be brought to this dark and narrow canyon where the huge wall was proposed to rise and the river to follow a new, ordered path.

Despite the lack of infrastructure to support the workforce, political pressure soon forced Frank Crowe's hand. In order to alleviate the hardships caused by the Great Depression in some small measure, President

Hoover urged the building of the dam to begin immediately. The town the workers were meant to inhabit had not been built; there was no sanitation, no canteen, just endless desert and the scruffy residents of Ragtown camped by the river.

To add to Crowe's difficulties, the Six Companies' contract with the government included completion dates for certain phases of the work. If these dates were not met, large penalties would be imposed that would eat into their profits. The first stage, before work could begin on constructing the dam itself, was to divert the Colorado River. Because of the narrowness of the canyon, this could be accomplished only by blasting tunnels through the canyon walls. Four tunnels, each 56 feet wide and three-fourths of a mile long, had to be excavated through the dense, black volcanic rock.

Crowe knew that the Colorado could not be diverted into these tunnels when it was in flood, carrying the snowmelt from the Rockies in late spring and early summer, and this created a problem. The government had set a completion date for the river diversion of October 1, 1933. If this date was missed, a steep penalty of $3,000 per day was to be paid by Six Companies. So Crowe had no choice but to divert the river into the tunnels earlier, during its low-water season in the winter of 1932. If this time were missed, a year would be lost before there was another chance to redirect the river, during which crippling fines would mount up. Crowe had, in effect, just 18 months to coax the heretofore untameable Colorado into three miles of tunneling, which was yet to be created in the sheer canyon walls.

IN LATE SPRING 1931, Crowe was a man under pressure as he began an all-out assault on Black Canyon. First, a railway between Las Vegas and the dam site was built, to bring supplies to the site, and roadways were blasted into the base of the ravine. Meanwhile, carpenters constructed wooden bunkhouses for the single men on the steeply sloping side of the gorge; and a few miles away, preparations were under way for a new town, Boulder City, which would provide modest houses for the married men with families. Men were hired: miners; electricians; so-called powder monkeys, who could handle explosives; and, arguably the most daring of all, the "high scalers," who had to brave the 700-foot height of the sheer cliffs while loaded down with tools and dynamite to prepare the canyon walls.

A high scaler trusted his life to a rope, which was attached to a steel eye-bolt driven into the rock at the top of the canyon, while he swung out hundreds of feet above the ground. These ropes, though, could fray or be cut by accident or break under strain. The scalers' aim was to hollow out the hard rock and trim the cliff face of all loose stone and rocky outcrops. The walls of the canyon had to be made sheer, smooth, and firm so that the massive curving wall of the dam could fit snugly into the sides of the canyon at its narrowest point. The design made use of the walls of the canyon as abutments, to help to absorb the immense weight of water trapped by the dam. Thus, there could be no faults, cracks, or blemishes where the concrete met the Colorado River rock; it would compromise the safety of the dam if water leaked at this point.

By May 1931, miners were ready to begin excavation on the four diversion tunnels, two on either side of the river. Although work was carried out in three shifts across 24 hours, progress was slow. The men labored alongside one another, drilling deep into the rock. Dynamite cartridges were then packed into drill holes and loaded with primers, which, when the area was cleared of men, would be detonated by an electric charge. At each charge, the ground trembled and the tunnel roared and spewed out thick clouds of smoke and the acrid smell of explosives. The noise from the machinery was deafening and was made worse by the thunderous roar of dynamite explosions as they echoed around the canyon. Orders to clear the site frequently went unheard amid the awful noise, so workers, unaware of an upcoming detonation, often suffered injury or were killed.

During the summer, the heat became a major problem as temperatures soared to an unbearable 120 degrees. Men working in the tunnels were subjected to even higher temperatures, sometimes more than 130 degrees. Drinking water from the river was pumped into large uncovered tanks, which in the searing sunlight were soon alive with bacteria. Amoebic dysentery spread fast, and the basic sanitary arrangements did little to alleviate the men's suffering. One resident described an "open toilet with three open seats, and a long line of waiting men appears constantly before these inadequate facilities . . . a physical disorder among the men made this an almost unbearable problem."

The summer of 1931 proved to be one of the hottest on record, and heatstroke began to take its toll. With little warning men working hard in the high temperatures would suddenly have a convulsive fit, followed by vomiting, and then collapse unconscious. It was not unusual for body temperature to rise to over 110 degrees. There was no treatment, unless someone had water nearby and could throw it over the victim by the bucketful. It was an hour's drive to the hospital at Las Vegas, and those who made it arrived packed in ice. For most, the journey was too long. "They arrived in Las Vegas dead, bloated, and looking like they had been parboiled," recorded one local doctor.

In Ragtown, too, with no shade, the humidity from the river, and the burning sand underfoot, there was no relief. Day after day the heat wave continued. Sandstorms blown by hot winds added to the suffering as the grit worked its way into eyes and mouths. Most had sunburn, too. Children became listless and sick, and in the inescapable heat, numbers of women and children died. On one day alone, July 26, three women died. The heat settled on the camp like a stifling mantle with no relief even at night. Sleep, when it came, was just "a heavy sweating coma," recorded one worker. As for the dormitories, "mice in the inhuman quiet make noises like people marauding. They are a kind of wild jumping mice that astonish me by hopping on the cots and running over the faces of the sleeping men, who are so used to them that they brush them away like flies."

The depressing squalor that the workers and their families had to endure, combined with the often dangerous work practices, were slowly undermining the morale of the men. There were also an increasing number of avoidable accidents. Crowe, however, was working to a strict timetable. As pressure mounted, he was always in a hurry, and safe practices became an expensive luxury.

Into this troubled situation stepped Frank Anderson, a union representative for the Industrial Workers of the World (IWW). He had arrived at the site in May, and as conditions deteriorated he lost no time pointing out to the men that the working practices were dangerous. Most of the workers agreed that Six Companies was taking advantage of them, but few were prepared to fight for better conditions for fear of losing their jobs. Many of them earned at least $5 a day, and there was no limit to the amount of food they could eat in the canteen. They were also hesitant about getting involved with the IWW, which had a reputation for its aggressive approach to employers. This uneasy state of affairs, which had been smoldering for some weeks, erupted on the afternoon of August 7, when Six Companies announced to the shift in the tunnel that there would be a reduction in pay.

Very quickly agitators persuaded the men that strike action was the answer, and Frank Crowe was handed an ultimatum. The men wanted a guaranteed wage of between $5 and $6 a day; they also wanted pure drinking water, flush toilets, and a safety miner with first aid at each tunnel; and they demanded that all Nevada and Arizona safety laws be followed. Next day the men had their answer. Six Companies dismissed their complaints and rejected their wage demands outright. Crowe insisted that the firm was blameless for the accident rate and pointed out that records showed no accidents for July—although the Bureau of Reclamation figures showed that 11 men had died of heatstroke. He argued there had been a misunderstanding about the wage cut, that it applied only to "muckers" who were being transferred from their jobs in the tunnels to new work outside, since machines could now take over clearing all the waste. Frank Crowe reminded them that a queue of hungry men willing to work stretched for a block and half in Las Vegas. So, apart from a few basic crew members, the men were all dismissed and told to take their belongings off the site. The message was clear: Six Companies would not be held to ransom.

Within a week the dispute had blown over. Most of the older and more reliable men were rehired, troublemakers were excluded, and Six Companies fulfilled some of the men's requests. Water coolers appeared, and men were hired especially to take water around to the workers on site. By September, the first families were leaving the squalor of Ragtown and moving into the company's housing in Boulder City, which, although very basic, were at least proper dwellings with showers and flush toilets.

Even with the strike over, for Crowe, as he embarked on the task of redirecting the ancient path of the river and forcing its wayward torrent through tunnels in the canyon walls, the big question was, how long would it take? He was a man for whom time had a special significance. He worked according to red rings on calendars, marking dates when penalties would begin to be assessed. With Six Companies bosses harassing him, and the human frailty of the workforce ever present in his mind, his job became a continual balancing act, with the whip of time constantly at his back. His tall figure was seen somewhere on the site, day or night, urging the men along. "Hurry up Crowe," the men called him behind his back.

The success of diverting the river successfully in the allotted time began to look doubtful. The 56-foot-diameter tunnels were so large they were impossible to dig out of the dense volcanic rock as a single section;

hence they had to be subdivided into smaller sections. The first area worked on was the "header" section, an area 12 feet by 12 feet in the very top center of the tunnel. When this was excavated, the work could progress to the wing sections on either side; then, in turn, the huge bulk of the tunnel, an area of 50 feet by 30 feet called the "bench" section, was tackled. When the bulk of the tunnel was dug, the bottom 12 feet of the invert section was excavated, leaving an enormous gaping 56-foot hole running into the dark belly of the rock. Four of these tunnels were needed, three-fourths of a mile long, all to be finished in 18 months.

The work was done in orderly fashion, but it was slow going until Crowe's valued assistant and second in command devised a plan to speed things up. Bernard "Woody" Williams designed a rig that was built around an old World War I chain drive truck. This enormous top-heavy contraption carried four platforms, which allowed two rows of miners—30 drillers in all—to work simultaneously. The other two platforms carried the drill steel. This extraordinarily cumbersome vehicle could be driven into the tunnel, backed right up against the wall and, when the drilling was done, removed equally quickly, saving much time. Each time the "Jumbo" was driven into position, more than 120 holes were drilled and packed with 2,000 pounds of dynamite, which advanced the digging by about 15 feet; and with almost 1,500 men employed in the four tunnels, up to 250 feet of tunnel could be excavated each day. It made the difference that Crowe so badly needed.

Moving the mountains of waste created by this excavation was soon turned into a fine art. Inside the tunnels, Caterpillar tractors shoved the debris into piles. Long queues of trucks were continually waiting to be filled with the 16,000 cubic yards of waste that was removed every day. Huge 100-ton electric shovels very quickly filled the waiting trucks, which then raced up to the disposal grounds a couple of miles away. To save more time and eliminate the need for turning space, the drivers reversed the empty trucks back down the twisty, sheer-sided canyon road at speed, all the while looking as though they were ready to jump off should they misjudge the edge of the steep-sided track.

The importance of speed was paramount in everybody's mind, and it turned men into automatons with no time to spare, all concentration devoted to the next task. One day, one of the men signaling the trucks forward, stationed outside the tunnel, happened to look up and was horrified to see a high scaler fall to his death. When he told the foreman that a man was lying dead near the trucks, the foreman, interested only in keeping the line of trucks moving replied: "What are you going to do with all these *blankety-blank* trucks? Eat 'em? Get the trucks moving! He can't hurt anybody."

As the work went faster and became increasingly mechanized, conditions for the men deteriorated further. Apart from heat and noise, the atmosphere was heavy with dust, the smell of explosives overpowering, and the air thick with exhaust fumes from the endless line of trucks clearing away rubble. The poisonous fumes in particular affected the men's health. "Carbon monoxide was knocking the men down, making them very, very ill," wrote one observer. "Yet when they were taken to the hospital, the doctors would say, 'It's pneumonia!' [The doctors] would not say, 'It's carbon monoxide poisoning and Six Companies is responsible.' " This state of affairs did not, however, escape the attention of A. J. Stinson, the Nevada inspector of mines, who ordered the company to convert all its petrol-powered vehicles to electricity. So, still running the race against time, Crowe again faced the prospect that work could be brought to a halt and that Six Companies would have to pay a bill for conversion costs to the tune of $1.5 million dollars.

And although he was making progress during the autumn of 1931 mining the canyon walls, again there was growing concern that the price being paid by the workforce was far too high. And it was just at this point, while Crowe was struggling with the endless conflicting demands of such a vast undertaking, that the Colorado River unleashed its power.

—

THE RIVER HAD been peaceful for months, lazily lapping the banks, but unseasonal torrential rain had fallen for days hundreds of miles to the north, filling the streams and rivers. As the Colorado traveled south, it gathered momentum, until it finally fell, engorged with mud and debris, into the canyon where the men were working. Roaring, it drowned out sounds from the workforce, its torrent rushing headlong through the canyon. Forcibly restrained where the canyon narrowed, the wall of water climbed ever higher up the black cliffs, threatening to wash away trucks and machinery. The workshops on the recently made canyon road were submerged and precious tools were swept away, along with vehicles, which were taken for an unexpected ride. Still the water rose, and if it reached the tunnels, lives could be in danger.

Fortunately, after having given an impressive performance as it lapped near the tunnels, the river fell back, leaving everything it did not take with it covered in layers of oozing slimy mud that took days to clear. No one had died and damage was limited, but this episode had cost time. Crowe hurried to get all the men back to work and pressured them to make up for lost time.

Tunnel superintendents would not allow the men to slacken; one famous hard case was, according to a worker's account, "so tough he could bite a nail in two and would fire a man for even looking like he was going to slow down." The tunnels with their explosives, electrical equipment, pools of water, wires, detonators, diggers, the poisonous cocktail of fumes, and men working against the clock had all the ingredients for tragedy. The local paper in Las Vegas frequently reported fatalities, usually involving the use of dynamite. Sometimes, the dynamite packed in the drill holes did not always explode, as a 23-year-old oiler, Howard Cornelius, discovered when his shovel struck a missed hole. Another worker, Gus Enberg, suffered a similar fate; his left arm was blown off, his right arm mangled, and his left leg shattered. Like Cornelius, he soon died of his injuries. On a different occasion, a worker was crushed between two dump trucks, fracturing his skull and crushing both his legs. In shock and hemorrhaging, he died soon after.

One night worker, Marion Allen, who arrived in November 1931, kept a record of his time at the dam. He was still new to the site when he was asked to take a message to the man in charge of the drilling Jumbo in Tunnel 3. He ran off, found what he thought was the right tunnel, and as he walked deep inside, was somewhat surprised to be met with an eerie silence and no sign of any workers. "I had gone quite a way when I had a funny feeling that something was wrong," he recalled. The sudden realization of what this unaccustomed deathly quiet meant prompted him to turn and run for his life. Seconds later, the shock wave from an explosion caught him and threw him down amongst a shower of stones. Allen was lucky and escaped relatively unscathed.

Against mounting criticism of their casual approach to safety, Six Companies decided to put up a fight against A.J. Stinson, the Nevada inspector of mines, by applying for an injunction to stop him from interfering with its use of petrol-powered trucks in the tunnels. It would take six months before the case came to court, during which work continued at the usual furious pace to drive the vast chasm through the side of the canyon. Until the case was heard, the petrol-powered trucks continued to be used, filling the tunnels with a heavy pall of choking and poisonous fumes, which hung like a blue mist at the mouths of the tunnels.

One of the advantages in choosing Black Canyon as the site of the dam lay in the particular quality of the rock, mostly andesite and basalt, which was easy to drill, and being free from fault lines needed no support during excavation. This fact, coupled with the enormous drive that Crowe brought to the project, meant that progress on the tunnels was so well advanced by the end of January 1932 that the night team working on Tunnel 3 going north broke through the rock to meet with the upstream crew coming south. Then, just a few days later, the same breakthrough occurred in Tunnel 2. All that remained was to excavate the lower invert sections of these tunnels.

Two weeks later, on February 9, the volatile Colorado River struck back. Work was halted again as the river, swollen with snowmelt, found its ancient strength and came roaring through the canyon. The men raced to remove equipment from the tunnels before the trestle bridge over the river was swept away. Three days later, the water climbed even higher, rising like a great wall—17 feet in just over three hours. Having evacuated the tunnels just in time, the workers watched in awe as the turbulent surge of water rushed at 50 miles an hour through the empty tunnels inside the canyon walls, destroying everything in its path. Crowe feared that everything would be washed out of the tunnels. When the river returned to normal, three feet of syrupy mud covered everything and took a week to clean up.

In spite of the floods, by the late spring of 1932, all four tunnels had been excavated, and by mid-March the first concrete was poured into the lining molds of Tunnel 3. All the tunnels were to be lined with three feet of concrete to allow the diverted river to flow smoothly. Each tunnel was concreted in three sections, starting with the invert, or bottom, section. Steel plates shaped to the curve of the tunnel, the forms, were held in place by jacks while concrete was poured into the space between the form and the rock. The concrete was poured through a number of carefully placed chutes starting at the bottom and working upwards. The overhead section was worked in exactly the same way, but the forms were held in position by vertical jacks, while inclined jacks held the sides. When the concrete was hard, the forms were moved onto a steel framework, or form carriage, 80 feet long and 40 feet high, to the next area. All cracks and blemishes in the new concrete were then grouted to ensure that the tunnel surface was completely smooth. This was essential to ensure there was no point of weakness through which the water might invade, causing the lining to break apart.

The case brought by Six Companies against A.J. Stinson and the state of Nevada was heard at the end of April 1932. Six Companies argued that it had invested $300,000 in petrol-powered vehicles and that the cost of conversion would be prohibitive. They explained that tunnel workers were not in danger from poisonous fumes, as there were plenty of air currents provided by fans and blowers. The company also questioned the legality of a state interfering in a federal government project.

Many were outraged by the court proceedings. Murl Emery, who ran the ferry and the local store, remembered vividly the night he had been called to help in a disaster. "They were hauling men out of those tunnels like cordwood. They had been gassed. I don't know how many was gassed. They were sick, real sick; they weren't dead. They were real sick from being gassed working in that tunnel along with the running trucks. And that die hard Frank Crowe in the courthouse, swore that to the best of his knowledge, there never had been men gassed working underground."

There is no doubt that many men suffered poor health and that some died from lung-related illnesses after working in the tunnels. But compensation was strictly allocated to the families of those who were actually killed on a job. "We believe that a great injustice is being perpetrated against the workers at Boulder Dam," wrote the Central Labor Union of Nevada to the chairman of the investigating committee on reclamation, "in the general lowering of working and living conditions on a project directly under the supervision of the government at this time of economic depression and unemployment. Labor at Boulder Dam has no voice in the settling of wages, hours of labor, working conditions, safety or living conditions."

The judges, however, after weighing the matter, and noting that the tunnels were almost finished, decided there was little point in converting the trucks and found for the company, issuing an injunction against the Nevada state mine inspector. Not a single vehicle was ever converted and, by late summer of 1932, the diversion tunnels were completed, well ahead of schedule. The highly dangerous task of taming the wild Colorado River was about to begin.

THE UNGLAMOROUS NAME of Boulder City given to the government reservation established for the dam workers could foster no illusions of grandeur or summon up much prospect of fun. The houses were so small they were more like rabbit hutches; all identical, on identical plots, street upon street, they looked like a lesson in perspective as they stretched out to the horizon. The city for the workers was to be a model township, protected as part of the Boulder Canyon Project Federal Reservation, which covered an area of 144 square miles. The Department of the Interior appointed one man to manage the growing city and ensure that it developed in an all-American and clean-living way. Sixty-nine-year-old Sims Ely—a retired banker, newspaper man, government official, who looked like one of the original Puritan fathers straight off the boat—was their man.

Everything that happened in the city was to be under Ely's jurisdiction, and he did not miss much. He was head of police, head of justice, and in charge of new businesses to see that they developed in accordance with good, old-fashioned family values. With equal justice, he settled disputes between man and wife and between business partners like some austere twentieth-century all-seeing Solomon. Marion Allen, an early citizen, thought that "one of the reasons Boulder City was such a nice place to live was due to the managership of Sims Ely. There was a constant stream of people to his office with applications for a dance hall, a gambling casino and every other business one could think of to separate a man from his money."

Temptation, however, was nearby. Twenty miles away, Las Vegas was essentially a nonstop party of prostitution, gambling, and heavy drinking. Ely made sure Boulder City would be free of such temptation. There would be no liquor sold in bars or shops; and all of the other delights that seem so essential after a few drinks were prohibited as well. Wages were paid fortnightly, and those intent on escaping the enforced virtue of Boulder City drove to Las Vegas, where the drink flowed faster than the Colorado and the gambling promised easy money, which could then be spent in the red-light district of Block 16, which beckoned with a blowsy and exotic charm.

The high scalers were amongst the elite of the workforce at the dam. After an eight-hour shift spent suspended from a rope hundreds of feet above the silver river, blasting and dynamiting the canyon sides, they considered it normal to collect their wages and spend a bawdy night in Las Vegas, drinking and whoring until daylight. But a night of high living was brought to an abrupt end at the reservation checkpoint, where those demonstrably drunk were not admitted—and might never be, if Sims Ely were involved. Ely was in no doubt that alcohol, dynamite, and workers—high scalers in particular—were not a good mixture.

The high scalers were men for whom risk held a special charm. As if to show a brazen disregard for their hazardous occupation, they would often entertain their fellow workers when the foreman was not around. Stripped to the waist, the high scalers would become the dazzling stars of a unique trapeze act, with the Colorado River miniaturized far below, as they competed to see who could swing out the farthest. The *Las Vegas Age* described the antics of one high scaler, Louis Fagan, in June 1932. Fagan, who was known as the "Human Pendulum," regularly performed the astonishing feat of delivering fellow workers from place to place around the cliff face. "The man who was being transferred . . . locks his legs around the waist of Fagan. Both men get a good grip on the rope and Fagan begins his daredevilry by bucking off at a tangent, which swings the human pendulum one hundred feet out into space and around the projection of the cliff to where they make the high-line cable coming up from the lower portals." When this was done, Fagan then went back for the next man. At the end of the shift, they all had to be brought back.

Certainly, one government inspector, Burl Rutledge, had reason to be grateful for the derring-do and skill of the high scalers. When examining work at the top of the canyon, he leaned just a little too far, lost his balance, and began the terrifying 700-foot fall to the bottom. Working just below him was high scaler, Oliver Cowan, who on hearing Rutledge's desperate scream, immediately swung out and caught him by the trousers. Another high scaler joined Cowan and the pair just managed to hold him against the cliff until a rope could be lowered.

Nonetheless, horrific accidents were not uncommon as high scalers picked their way through the thick skeins of electric lines and air hoses and tried to dodge the falling rocks and tools. Often a shout from above was the only brief warning of a falling object. One high scaler, Jack Russell, did not hear the warning as men were lowering the drill steel down the cliff face; a steel rod came loose and struck him on the head. Without uttering a sound, Russell fell, poleaxed, and his comrades could only watch as he was thrown off each rocky outcrop to land at last in an unrecognizable heap at the bottom of the canyon. The report of the accident described how "brains and blood with pieces of torn flesh and tattered clothing were scattered over the jagged rock." The fact that he landed within the boundary of Arizona may have been some small consolation to his widow, as Arizona paid higher rates of compensation than Nevada. Most men hoped that if they died, they would die in Arizona.

One solution to the dangers of falling objects from above was the ingenious invention of the "hard hat." The men put two baseball caps together, one brim at the front, the other turned to face the back. It was then dunked in tar several times and, when dry, became so hard that it saved many men from fractured skulls. The company liked the invention so much that they ordered hard hats for everybody. Even so, Marion Allen, who worked up high for about a year, confessed, "I believe I was scared the entire time."

AS THE QUIETER weather of autumn 1932 spread like a balm over the canyon, and the burning summer skies turned to softer colors, the work of directing the river from its ancient course began. The diversion tunnels were ready and Frank Crowe was anxious to redirect the Colorado while it was at its seasonal low. First, he had to construct two vast temporary cofferdams, one high enough upstream to force the river into the tunnels, and one downstream, to stop the diverted water from flooding back upstream. The aim was to create approximately a mile of dry riverbed so the men could prepare the site for the 660-foot base of the dam. It was a huge feat, which left plenty of scope for things to go wrong.

The process began during October, when a long line of trucks tipped dirt and rock scree into the Colorado from the Nevada bank, just below the entry tunnels for the river. Tractors bulldozed the endless pile to form a bank, which reached out into the middle of the river, where it turned, ran 500 feet downriver, and then turned once more to join the Nevada bank, forming a cofferdam. This was then pumped dry and would enable an early start to be made on the main upper cofferdam wall. It also had the effect of pushing the river nearer the Arizona bank, making it easier for the river to enter Tunnel 4.

In mid-November, dynamite blasted away the temporary barricade in front of Tunnel 4. Trucks began tipping rock waste from a trestle bridge into the river to form a temporary barrier; it took all day and all night as the trucks in a long queue continuously disgorged the mud and rock. Slowly, as the massive barrier grew, the river rose, swollen and dangerous within the narrow canyon walls, fighting the obstacle that lay in its path. At last the barrier was higher than the turbulent swirling water and the river was forced to sing a new song as it roared through the perfect concrete symmetry of the tunnel. Those few who had come to watch on that raw November morning stood in awe, as though witnessing some miracle of the Old Testament while the water was swallowed up by the black rock. For Frank Crowe it was a great triumph. Man had won against the river, diverting it from the ancient course it had traveled for 12 million years.

There was still work to be done, though. The mountain of concrete that would hold back the vast reservoir could not be built until two cofferdams completely sealed off the old riverbed to leave a dry site. It was another race against time since the upstream cofferdam had to be strong enough to withstand the wildness of the river when it carried snowmelt, and the volume of water could increase fourfold. Once again, the endless queue of lorries dumped their rocky waste, and bulldozers set to work 500 feet downstream from the diversion inlets building a barricade from bank to bank 100 feet high. To ensure the cofferdam would not let water ooze through where it joined the canyon wall, reinforced concrete married the two surfaces. When finished, the mighty barrier was 750 feet wide at the base, with the face that met the river covered in 3 feet of concrete. The downstream cofferdam, which was to stop water circulating back to the dam site, was smaller, around 200 feet thick at the base and over 50 feet high.

The riverbed was pumped dry and centuries of silt were removed as men walked along the ghostly line where the river had left its watery mark on the canyon walls. Acres of mud and great swaths of stones lay in waves against the bank, while slime covered the river bottom. Over 20,000 cubic yards of silt was removed every day for many weeks, finally revealing bedrock 40 feet below river level. In the center of the canyon, the river had scoured out a chasm 80 feet deep that lay full of black brackish water. By early summer 1933, the bottom had been picked clean. Pipes drained any rising water, springs were sealed, fissures were blocked, and the surface of the river bottom was smooth and dry. Finally, on June 6, the first delivery of concrete was made. A bottom-drop bucket carrying eight cubic yards of concrete was placed into a prepared framework in the deepest part of the canyon, and the job of building the massive wall of concrete began.

From the very beginning, when estimates had been put together for the dam, Crowe had insisted on assessing high costs for all the early preparation work: the roads, tunnels, river diversion, and cofferdams. His estimates on the later stages of the construction were much lower. Six Companies directors had been full of admiration for Crowe as government money poured in for the first stages, which were completed ahead of schedule. These payments had covered their initial $5 million outlay and had started to bring in big profits, too. Now they had to build the huge concrete dam wall itself, and potential profits were suddenly much less promising.

Crowe was not a gambling man, but he *was* willing to gamble on himself. If there was a more effective way of doing a job, he had either invented it or knew about it. He was convinced he would save time building the dam wall by following his unique plan for a system of overhead cables, which would carry men and materials quickly to anywhere on the dam site. To win the contract, he had persuaded Six Companies to slash the price of the 3.5 million cubic yards of concrete needed by over 30 percent. He had assured Six Companies directors that even with this low quote he would complete the work ahead of schedule and still make good profits. This, of course, assumed that nothing would go wrong at this stage to endanger the slim margins.

Crowe was 18 months ahead of schedule and meant to keep things that way. In order to continue to get work accomplished at the fastest possible pace, an element of competition was fostered among the workers. Throughout the soaring summer temperatures of 1933, the concrete colossus rose, a modern monument of incomprehensible complexity, its myriad columns—all at different levels and angles—shimmering in the heat. Teams were pitted against each other and bonuses were paid for fast, efficient work. But speed was often a killer, as the big posters around the site reminded everyone: "Death is so permanent."

The series of columns that formed the framework for pouring the concrete were all interlocked both horizontally and vertically in an intricate mesh that gave immense strength. They rose in a series of separate wooden blocks, or forms, up to 60 feet square and 5 feet deep: high windowless rooms of solid concrete. Once the concrete had set, the form could be taken apart and rebuilt on top, ready for the next pour. Keeping the concrete cool was the greatest challenge. The chemical reaction caused when concrete sets generates heat, which could lead to fractures and cracks. The Bureau of Reclamation calculated that if the concrete were delivered in one continuous pour, the temperature would rise by 40 degrees and take 120 years to cool.

Consequently, to ensure adequate cooling, there was a four-day gap between placement of the five-foot sections, and no more that seven sections could be placed on the same column in one month. In addition, cold water from a refrigeration plant was circulated through pipes in the concrete. After each load of concrete, a team of men in rubber boots tamped the surface level, spreading it into place along the edges. When the concrete section had hardened, it was thoroughly cleaned with pressurized water, which left a rough surface for the next five-foot pour. After 24 hours, pipes for the refrigerated water were laid, and the wooden casing was stripped away and reset on another block for the next pour.

There were two mixing plants churning out concrete at such a rate they could fill up to 35 rail trucks each day. The wet concrete was delivered swiftly to the dam site through Frank Crowe's ingenious system of overhead cables. This cat's cradle of overhead wires could transport men and materials over 1,200 feet each minute anywhere on the dam site.

There were five cableways crossing the canyon, each capable of carrying a 20-ton load, not to mention four smaller cableways and the government's cableway. These connected to towers, some up to 90-feet high, on either side of the river, which were mounted on tracks so they could travel along the canyon as the need arose. Additional pulleys and cables attached to the main cables allowed materials to be moved across the canyon and up and down so that every part of the site was within reach. Each cable section was managed by an operator, perched precariously in a hut on a platform high out on the canyon walls with an eagle's eye view of the whole endeavor. The system was so efficient that Marion Allen recalled the day in March 1934 when the entire team of workers poured 10,401 cubic yards of concrete. As they were congratulating themselves on their success, the foreman came up, looking stern and shaking his head. "And you know it wasn't a very good day for pouring concrete either," he said harshly.

Despite the legendary skill of the cable operators, danger was a constant. Early one morning, Allen and some other workers were being transported across the canyon in a skip—a large open wooden crate for 50 men that would hang from a cable moved by the pulleys—when it started spinning. The skip was 400 or 500 hundred feet out, "when the cable became so twisted it started throwing sparks and looked as if it was coming apart and there we were with two or three hundred feet of open space below us and nothing to stop us if the cable broke." Everyone screamed for the skip to be stopped, but when it was, the cable continued twisting. After some hours in the hot sun, with the crate dangling over the canyon, a rigger was sent out to try and unwind the cable. "Even as scared as we were, I still worried about the rigger, clinging all the way out," said Allen, "with nothing to hang on to but the greasy cables." The rigger, however, could do nothing except suggest they all move to one side to see if that would correct the cable, which did not work. Finally, by midafternoon another skip was sent out along the cable, to which the men transferred in midair; and from their new, safe skip, they unwound the cable of the first one.

Not everyone, of course, was so lucky. One site worker, Jack Egan, by some strange quirk of fate, decided at the last minute not to accompany his friend Kenneth Wilson for an easy ride from the bottom to the top of the canyon. They had been preparing a piece of machinery that was to be lifted to the top of the canyon by cable, but just before it left, Egan had jumped off. Four hundred feet up, the cable broke, sending Wilson and the machinery plunging to earth with such force that the machine was half buried in the soil. As Egan stood and watched in horror, he saw his friend wave good-bye to him as he fell.

Courage was the unspoken but necessary qualification for anyone who worked at Hoover Dam and courage had been in plentiful supply amongst the men who had gouged out the four three-quarter-mile-long tunnels in temperatures of 130 degrees, and in an atmosphere heavily laced with the lethal exhaust of a hundred lorries. But six of these workers whose health had been ruined now decided to sue Six Companies for carbon monoxide poisoning. Their lawyer, Harry Austin, was claiming at least $75,000 for each of them. Many more workers watched the proceedings with interest. Thousands had been suffering health problems after working in the dreadful conditions and wondered what their chances were of making what looked like an easy $75,000 from a very rich company. The hearing was set for October 1933.

ONCE THE COLORADO River was dammed, its water would be carefully regulated, instead of running wild as it had in former days. Some of the water would flow through turbines to generate electricity; the rest would pass into the riverbed. Provisions had to be made to handle excess water in times of flood, so vast spillways were sited at the top of the dam to carry any water that overflowed the spillway weirs to the river channel below. They were 700 feet long and built on either side of the dam. At the bottom of each spillway, a 70-foot-wide concrete tunnel, known as the "glory hole," plunged at a sharp angle of 45 degrees through the canyon walls to remove any excess water. The tunnel dived through the rock, narrowing to 50 feet to join the river-diversion tunnels below, taking the overflow water downstream.

The first attempts at lining these massive tunnels with concrete proved difficult. The concrete was sent down in large buckets by cable car, but on arrival it had invariably set firm. Delivering it by pipeline was tried, too, but to no avail; in the searing temperatures of midsummer 1933, it simply stuck to the pipe and set solid, eventually blocking it completely. Living up to his name as an innovator, Crowe finally found the solution: a portable agitator mounted on a steel casing kept the concrete from setting, and work soon returned to its usual furious pace.

In the ovenlike temperatures in the steeply sloping tunnel, working conditions were, essentially, a stage set for disasters. According to Allen, on one occasion, four carpenters on the night shift, having removed the bolts from a form that was to be used higher up in the tunnel, stood on the form just a little too long as the cables pulled it slowly upwards. Without warning, part of the form became detached and started a downward journey, with three of the carpenters still on it. They clung on precariously as the form gathered speed. It left the glory hole tunnel, dived into the dark river-diversion tunnel, where it was swept along until it emerged minutes later into the Colorado, delivering the three carpenters, suffering many broken bones, barely alive on a sandbank.

With 5,000 men employed at the canyon, the dam was growing daily as the massive wall of concrete took each inexorable five-foot step higher. Then, one terrible incident occurred: a form collapsed as concrete was placed into a five-foot section, next to the central shaft, which housed the water refrigeration pipes. Two men were working in the central shaft when all the concrete that was in the five-foot section began to pour down on top of them. One of them heard the sounds of the oncoming deluge and scrambled to safety, but his friend, W. A. Jameson, was enveloped by 100 tons of falling concrete, which continued down another 100 feet. He was entombed in the rapidly setting concrete. Frantic efforts were made to reach him but his body was not recovered until the next day. It was the sudden and unforeseeable nature of such accidents that left such a disturbing impression on the men.

Meanwhile, Six Companies had settled on a strategy to deal with the impending lawsuits over carbon monoxide poisoning. Should one or more of the plaintiffs win, the floodgates would be open to countless more with similar claims. Company management brought in the police chief at Boulder City, Bud "Butcher" Bodell, who specialized in detective work and maintained a band of informants. As head of Six Companies security, his task was to see if any evidence could be found that would challenge the six claimants' pleas of ruined health.

Bodell soon established that at least one of the claimants, Ed Kraus, had an extremely interesting lifestyle, which featured a steady diet of all-night dancing and drinking in Las Vegas. He also had a police record for theft. The company investigators soon came to the conclusion that if they gave Kraus a little more rope, he might hang himself, and the attorney, Harry Austin, would be obliged to drop the case—and possibly the others, too. To ensure their outcome, Bodell employed a shady character named John Moretti, who was instructed to find a way of directly incriminating Kraus. That summer, Kraus fell into their trap, carousing with women and engaging in petty criminal activities, which ended with assault and robbery of a retired miner, for which he was charged. Moretti also uncovered Kraus's complicated plan to blackmail important Las Vegas citizens with a little help from an extremely attractive young woman.

By a convenient coincidence, the trial of Ed Kraus was to be heard only two weeks prior to his criminal trial. At the claims trial in October, Kraus told of his symptoms: his lost sex drive and the illness that befell him when he worked in the tunnels as a driver, where sometimes the air was so thick from the fumes of waiting lorries that it was impossible to see the electric lights. But Six Companies had employed lawyer Jerome White, who, with his razor-sharp mind, did his utmost to demolish the character of Kraus. With the help of half the lowlife of the West, White also threatened to give evidence that would prove that far from losing his libido, Kraus in fact had been enjoying all the heady pleasures that Las Vegas had to offer, notably with someone called Merle.

The disclosures of the various exploits of Ed Kraus held the town transfixed, and the jury became so bemused that a verdict was beyond them, whereupon the judge announced a mistrial. At his criminal trial that followed, it became only too plain that Six Companies had been busy "building" their case. The robbed miner was revealed to be a friend of Bud Bodell, and the jury found Ed Kraus not guilty of assault and robbery.

In stark contrast, the next claims hearing was heard in ordered calm as Harry Austin made his plea for a clearly very sick man, Jack Norman. As witnesses gave evidence, it looked as though there would be an obvious victory for the claimant. When the jury members retired to consider their verdict, however, Austin discovered through a drunken colleague of Bud Bodell that nine of the jurors had received a payoff from Six Companies to bring in a verdict favorable to the company. Austin did his best to fight the corruption but could find no one willing to testify. According to one juror, "If I do, somebody will shoot me in the guts."

Austin, refusing to bow to the might of Six Companies, brought a further 48 cases to court, with claims amounting to nearly $5 million. The problem for Six Companies would not go away, and their ruthless handling of the situation had not dissuaded fair-minded men that real impairment to health had been suffered by those who worked in the tunnels.

Eventually, Six Companies accepted the inevitable and quietly settled all the cases out of court for undisclosed sums.

DURING 1934, WHILE the sheer wall of white concrete rose in Black Canyon, an enormous warren of tunnels hidden out of sight in the sides of the canyon were under construction to operate the dam. The complex system that drove the electricity turbines began in four intake towers, which were situated immediately behind the dam. The towers themselves were graceful columns of concrete rising nearly 400 feet from ledges that had been cut in the canyon walls nearly 250 feet above the old riverbed. They were designed to draw off water from the reservoir that would form behind the dam, Lake Mead, and deliver it through a series of pipes, called the penstocks, to the powerhouse. Each tower, 75 feet in diameter, was heavily reinforced with steel, and electricity-powered gates controlled the flow of water.

The water from the intake towers was to be delivered by means of a massive 30-foot-diameter steel pipe to a series of smaller pipes that directed the water to turbines. But getting the 30-foot steel pipe sections from the manufacturer in Ohio to the dam proved impossible. Each section of pipe was big enough to fit a three-bedroom house in it, and there wasn't a vehicle in the country able to transport such a large piece of equipment. The manufacturers had no alternative but to build a new steel plant near the dam. Even then delivery of the huge sections of pipe from the factory to the dam was a problem, until Six Companies provided a Caterpillar tractor to haul a trailer with the colossal 200-ton sections of pipe in tow. Then, it was no easy task to lower the unwieldy pipe down the canyon to the tunnel entrance by cable and maneuver it into position. Once in place, however, the sections were sealed together, creating an underground maze of shiny steel in the ancient cliffs.

By the summer of 1934, the colossal concrete wall of the dam was almost at its final height. The cement finishers were busy smoothing the sides, filling in irregularities, slowly erasing the telltale ridges every five feet—like the rings on trees—where the forms had been. On February 1, 1935, a great steel gate closed off the last diversion tunnel, and the river, which for centuries had run wild and free, was now humbled to follow a controlled and orderly process. Silently, slowly, Lake Mead began to fill behind the dam, submerging all the familiar rocks and tracks and signs of endeavor with a flat sheet of water reflecting the clouds. The countryside around began to benefit. A huge flood of water in early summer, which would formerly have deluged the Imperial Valley, was now controlled; the river's good behavior meant plentiful water in the season of drought that came later. The Colorado River was at last subdued to man's will.

With the end in sight, Six Companies faced a final showdown with the workers. Another accident and the handling of compensation for it highlighted the growing dissatisfaction of the dam workers, and led to a strike. One of the men, Eugene Schaver, fell at work in May 1935, and was knocked unconscious. His head injuries were so severe he soon died in the hospital. But according to the doctors, he had died of spinal meningitis. The dead man's colleagues knew that he had fallen at work, hit his head, and died of his injury, and that his family should be compensated by the company. A diagnosis of spinal meningitis would not bring compensation, and Schaver's angry fellow workers thought it most likely that Six Companies was trying to evade honoring its promises as it had in the past.

By now, a different atmosphere prevailed amongst the workforce, who were less inclined to kowtow to the employers. In 1935, the worst days of the Depression were over. President Roosevelt's New Deal, promoting government-funded construction projects, had helped increase employment opportunities and fair wages. The unions, which had struggled to survive in 1931, were now a growing force and had found a voice.

Workers at the dam were well aware that other government projects paid higher wages.

So when a few weeks after this accident Six Companies decided to increase each shift by half an hour with no extra pay, the men doing vital work on the powerhouse were ready for action. President Roosevelt was due to dedicate the dam officially in September, in two months' time, so the men felt they were in a strong position and went out on strike, demanding higher wages and a six-day week. It was high summer and most were glad to take a few days off from the endless grind of work. "The duration of the strike was one of the most pleasant times I can remember at Boulder City," recalled Marion Allen. "We had plenty of time to go to Lake Mead where we could swim or we picnicked almost every day."

But the dam was so close to completion by the summer of 1935 that Six Companies could afford to play a waiting game. They predicted the men would be back when the money ran out. They were right. With no advantages gained—except that the proposed extra half an hour's work was canceled—the men slowly gravitated back to work.

Two months later, on September 30, 1935, everything was ready for the official dedication by President Roosevelt. Boulder City came alive with a great crush of 12,000 people in festival mood who had come to see the opening of the dam. The day was fine and warm, and the colorful dresses of the women in the crowd made a floral garland around the severe lines of the dam. The president's car arrived, and as he stepped out in the bright sunlight, that familiar figure, a symbol for so much hope, stood surrounded by dignitaries.

The president took in the scene of the huge dam, beyond human scale in its vastness. "I came, I saw, and I was conquered," he said, his voice heard by millions more on radio. He could see that a modern miracle had transformed this desolate place. "Ten years ago, the place where we are gathered was an unpeopled, forbidding desert. The transformation wrought here is a twentieth-century marvel." He was clearly moved to see the dam in all its glowing pristine whiteness, like a temple to the modern age. "All these dimensions are superlative," he continued. "They represent and embody the accumulated engineering knowledge and experience of centuries; the energy, resourcefulness and zeal of the builders. . . . But especially we express our gratitude to thousands of workers who gave brain and brawn to this great work of construction."

Marion Allen listened to the president on that sunny day and felt an intrinsic part of the enormous adventure: "His voice is something I'll never forget, for it had some quality in it which could move mountains," he said. Perhaps he forgot for a moment that it was he and a whole town of his workmates who had actually moved the mountain. Allen's experiences and insights represent those of the many nameless men whose lives were lived in the shadow of the great dam. He had experienced it all: Ragtown, the early work to move the river through the walls of the canyon, the heat, the accidents, the huge concrete wall climbing up the canyon, the finished work now so perfect, hiding all the long years of effort. He stayed until the dam was completed and then went straight on to another big construction site, saying that he had "contracted a very common disease known as constructitis," for which "there is no known cure. Sometimes one thinks he is cured, only to have a relapse when he goes by fresh concrete or catches the smell of fresh sawdust from new lumber. Anyone with this affliction has to start construction of some kind, even if only to dig a hole and fill it up again."

For his part, Frank "Hurry up" Crowe collected a bonus of up to $350,000 and the respect of millions—not least from the president and Six Companies management. He had carried the responsibility for undertaking and completing the complex task, which he had done, to everyone's amazement, two years ahead of schedule. But, like Allen, Crowe was addicted to work. He was already organizing work for his next site, the Parker Dam, 150 miles away, as he was finishing Hoover Dam, and he built several more dams, including the acclaimed Shasta Dam in California. Eventually, the unrelenting pressure to which he had subjected himself for so many years demanded a price. In 1945, he had to refuse the offer to help reconstruct the U.S. Zone of Occupation in Germany. He died in 1946, at his farm in California, a huge body of work to his credit.

According to Bureau of Reclamation figures, 107 men lost their lives while working at Hoover Dam and many more carried the scars of their efforts. On December 20, 1935, a man called Patrick Tierney fell more than 300 feet from the top of an intake tower and drowned. He was the last man to die on the project. Ironically, his father, J. G. Tierney, had been one of the first to die at the site, exactly 13 years before.

The silenced streets of Ragtown lay lost somewhere under the slowly rising water, only a memory now. The wild Colorado River was tamed, and with the blessing of abundant water, the arid lands of the Southwest flourished like an opening flower. In 1936, when the first turbines were switched on, electricity began to light the growing desert towns; and Las Vegas continued to host its all-night party for the world, forever dazzling the night sky with its luminescence.

The benefit of the dam to the area was immeasurable. The previously empty deserts now supported a whole string of reservoirs, irrigation systems, canals, and aqueducts, turning the burning wastes from the Rocky Mountains to the Pacific into a promised land. The massive generators at the base of the dam can generate up to 3 million horsepower and provide electricity for three states. The desolate region that had once looked to the East Coast for its inspiration was transformed into a rich and brilliant land.

Arthur Powell Davis's original vision, which had haunted him as a young surveyor, was now a reality. The deep wide lake, holding so many riches, stretched as far as the eye could see and beyond, for 150 miles.

And the stunning white wall that held it all together was the glorious best that America could do in those dark days of the Depression. The Hoover Dam became a trophy for the nation. And although larger dams have since been built, not one could usurp the special position it holds as America's celebrated "damn big dam."

BIBLIOGRAPHY AND SOURCES

THE GREAT EASTERN

—

Primary Sources

Brunel University, Uxbridge: Brunel Collection (Photographs)

Family Records Centre (Census returns for 1851 and 1861)

House of Lords Record Office (Parliamentary Papers, Hansard)

Institution of Civil Engineers (Miscellaneous deposited papers of the Brunel family)

National Archives: Public Record Office (Admiralty papers, dockyard employees, and shipbuilding)

University of Bath Library: Hollingworth Collection (Copies of correspondence relating to the construction of the *Great Eastern*)

University of Bristol Library: Special Collections (Eastern Steam Navigation Company letter books, Brunel's private letter books, sketchbooks, calculation books, desk diaries, personal correspondence, and miscellanea)

Journals, Newspapers, and Periodicals

Chatham News

The *Engineer*

Household Words

Illustrated London News

Institute of Naval Architects, Transactions

Institution of Civil Engineers, Proceedings

Mariner's Mirror

Morning Chronicle

The *Observer*

Scientific American

Technology and Culture

The *Times*

Secondary Sources

Barry, P. *The Dockyards and the Private Shipyards of the Kingdom*. London: T. Danks, 1863.

Beaver, P. *The Big Ship: Brunel's* Great Eastern. *A Pictorial History*. London: Hugh Evelyn, 1969.

Booth, C., ed. *Labour and Life of the People*, 2 vols. London: Williams & Norgate, 1889, 1891.

———. *Labour and Life of the People*, 2nd ed., 10 vols. London: Macmillan & Co., 1892–97.

Brunel, I. *The Life of Isambard Kingdom Brunel, Civil Engineer*. London: Longmans & Co., 1870.

Buchanan, R.A. *Brunel: The Life and Times of Isambard Kingdom Brunel*. London: Hambledon, 2002.

———. "The First Voyage of the SS *Great Eastern*," *Proceedings of the International Committee for the History of Technology Symposium*. Vienna, 1991.

Dugan, J. *The Great Iron Ship*. London: Hamish Hamilton, 1953.

Dumpleton, B. and Miller, M. *Brunel's Three Ships*. Melksham: Colin Venton, 1974.

Emmerson, G.S. *The Greatest Iron Ship SS* Great Eastern. Newton Abbot: David & Charles, 1981.

———. *John Scott Russell: A Great Victorian Engineer and Naval Architect*. London: J. Murray, 1977.

Free, R.W. *Seven Years' Hard*. London: William Heinemann, 1904.

Golch, R.B., ed. *Mendelssohn and His Friends in Kensington: Letters from Fanny and Sophie Horsley Written 1833–6*. London: Oxford University Press, 1934.

The Great Eastern *ABC, or Big Ship Alphabet*. London, 1860.

The Great Eastern*'s Log: Containing Her First Transatlantic Voyage and All Particulars of Her American Visit by an Executive Officer*. London, 1860.

The Great Eastern *Steam Ship: A Description of Mr. Scott Russell's Great Ship, Now Building at Millwall, for the Eastern Steam Navigation Company*. London, 1857.

Griffiths, D., Lambert, A. and Walker, F. *Brunel's Ships*. London: in association with National Maritime Museum, 1999.

Hobbes, R.G. *Reminiscences of Seventy Years' Life, Travel and Adventure; Military and Civil, Scientific and Literary*, 2 vols. London: Elliot Stock, 1893.

Jackson, T. *Visitors Guide to the* Great Eastern. Holyhead: L. Jones, 1860.

Hutchinson, G. *The* Great Eastern: *A Concise History of This Remarkable Vessel*. Liverpool: G. Hutchinson, 1886.

Milbanke, M.C., Countess of Lovelace. *Ralph, Earl of Lovelace: A Memoir*. London: Christophers, 1920.

Mortimer, J.E. *History of the Boilermaker's Society*. London: Allen and Unwin, 1973.

Noble, C.N. *The Brunels, Father and Son*. London: Cobden-Sanderson, 1938.

Pudney, J.S. *Brunel and His World*. London: Thames & Hudson, 1974.

Pugsley, Sir A., ed. *The Works of Isambard Kingdom Brunel: An Engineering Appreciation*. Cambridge: Cambridge University Press, 1980.

Razzell, P.E. and Wainwright, R.W., eds. *The Victorian Working Class: Selections from Letters to the "Morning Chronicle."* London: Cass, 1973.

Rolt, L.T.C. *Isambard Kingdom Brunel*. Harmondsworth: Penguin, 1985.

Russell, J.S. *The Modern System of Naval Architecture*, 3 vols. London, 1864–65.

Vaughan, A. *Isambard Kingdom Brunel: Engineering Knight-Errant*. London: J. Murray, 1991.

Wilkins, G. *The Mystery of Providence: A Sermon by the Rev. George Wilkins Preached in the Sailor's Institute, Mercer's Street, Shadwell on Sunday Evening, February 5th 1860 on the Occasion of the Death of Captain Harrison of the*

"Great Eastern" with a Brief Account of His Early History, and a Narrative of the Melancholy Event. London, 1860.

THE BELL ROCK LIGHTHOUSE

Primary Sources

House of Lords Record Office (Parliamentary Papers, Hansard)

National Archives of Scotland (Northern Lighthouse Commission minute books)

National Archives: Public Record Office (Admiralty: Ships pay lists, Correspondence; Board of Trade: Minutes; Treasury: Correspondence)

National Library of Scotland (Stevenson Collection; Rennie letters)

Journals, Newspapers, and Periodicals

Dundee Magazine and Journal of the Times

Dundee, Perth and Cupar Advertiser

Dundee Weekly Advertiser and Angusshire Intelligencer

Edinburgh Advertiser

Edinburgh Evening Courant

Glasgow Advertiser

Glasgow Herald

Herald and Advertiser

Steele's Navy List

The *Times*

Secondary Sources

Allardyce, K. *Scotland's Edge Revisited*. Glasgow: HarperCollins, 1998.

Bathurst, B. *The Lighthouse Stevensons: The Extraordinary Story of the Building of the Scottish Lighthouses by the Ancestors of Robert Louis Stevenson*. London: HarperCollins, 1999.

Boucher, C.T.G. *John Rennie 1761–1821: The Life and Work of a Great Engineer.* Manchester: Manchester University Press, 1934.

Hay, G. *History of Arbroath to the Present Time, with Notices of the Neighbouring District,* 2nd ed. Arbroath: T. Buncle & Co., 1899.

Leslie, J. *Bright Lights: The Stevenson Engineers 1752–1971: Comprising Family Recollections.* Edinburgh: J. Leslie and R. Paxton, 1999.

Mair, C. *A Star for Seamen: The Stevenson Family of Engineers.* London: J. Murray, 1978.

Munro, R.W. *Scottish Lighthouses.* Stornoway: Thule, 1979.

Nicholson, C. *Rock Lighthouses of Great Britain: The End of an Era?* 2nd ed. Caithness: Whittles Publishing, 1995.

Skempton, A.W., ed. *John Smeaton FRS.* London: Thomas Telford, 1981; reprinted 1991.

Smeaton, J. *A Narrative of the Building and a Description of the Construction of the Eddyston Lighthouse with Stone,* 2nd ed. London: for G. Nichol, 1793.

Stevenson, A. *Biographical Sketch of the Late Robert Stevenson, FRSE, Civil Engineer.* London: W. Blackwood and Sons, 1861.

Stevenson, D. *Life of Robert Stevenson.* Edinburgh: Adam and Charles Black, 1878.

Stevenson, D.A., ed. *English Lighthouse Tours 1801, 1813, 1818: From the Diaries of Robert Stevenson and Drawings of Lighthouses.* London: Thomas Nelson, 1946.

———. *The World's Lighthouses Before 1820.* London: Oxford University Press, 1959.

Stevenson, R. *An Account of the Bell Rock Lighthouse.* Edinburgh: Constable, 1824.

Stevenson, R.L. *Records of a Family of Engineers.* London: Chatto and Windus, 1912.

Whittaker, I.G. *Off Scotland: A Comprehensive Record of Maritime and Aviation Losses in Scottish Waters.* Edinburgh: C-Anne Publishing, 1998.

Worth, M. *Sweat and Inspiration: Pioneers of the Industrial Age.* Stroud: Sutton Publishing, 1999.

THE LONDON SEWERS
—
Primary Sources

British Library (Metropolitan Sewers Commission: Reports)

Corporation of London Record Office (Guildhall Papers)

Family Records Centre (Census returns, 1851, 1861; Office of National Statistics, Indexes to birth, marriages, and deaths)

Institution of Civil Engineers (Bazalgette papers)

London Metropolitan Archives (Metropolitan Board of Works: Contracts and specifications, Main Drainage Committee, Thames Embankment Committee, Works Committee, General Committee, Plans, Printed Annual Reports; Metropolitan Sewers Commission: Minutes, Printed Reports, Correspondence)

National Archives: Public Record Office (General Board of Health Minutes and Correspondence)

Journals, Newspapers, and Periodicals

British Medical Journal

The *Builder*

The *Building News*

Cassell's Saturday Journal

City Press

The *Civil Engineer and Architect's Journal*

The *Economist*

Household Words

Illustrated London News

Institution of Civil Engineers, Proceedings

Journal of the Statistical Society

The *Lancet*

Marylebone Mercury

Medical Times and Gazette

The *Observer*

Punch

The *Times*

Secondary Sources

Aubuisson de Voisins, J.F. de, trans. Bennett, J. *A Treatise on Hydraulics for the Use of Engineers, Translated from the French and Adapted to the English Units of Measure*. Boston: Little, Brown & Co., 1852.

Bazalgette, J. *On the Main Drainage of London*, London: William Clowes & Sons, 1865.

Booth, C. ed. *Labour and Life of the People*, 2 vols. London: Williams & Norgate, 1889, 1891.

———. ed. *Labour and Life of the People*, 2nd ed., 10 vols. London: Macmillan & Co., 1892–97.

Booth, G.R. *The London Sewerage Question*, London: W.E. Painter, 1853.

Box, T. *Practical Hydraulics: A Series of Rules and Tables for the Use of Engineer*. London: E. & F.N. Spon, 1867.

Coleman, G.S. *Hydraulics Applied to Sewer Design*. London: C. Lockwood & Son, 1923.

Davis, A.C. *A Hundred Years of Portland Cement, 1824–1924*. London: Concrete Publications Ltd., 1924.

Downing, S. *The Elements of Practical Hydraulics for the Use of Students in Engineering*. London, 1855.

Ellis, R.H., ed., *The Case Books of Dr. John Snow*. London: Wellcome Institute for the History of Medicine, 1994.

Eyler, J.M. *Victorian Social Medicine: The Ideas and Methods of William Farr*. Baltimore, London: Johns Hopkins University Press, 1979.

Farr, W. *Report on the Mortality of Cholera in England 1848–9*. London, 1852.

Finer, S.E. *The Life and Times of Sir Edwin Chadwick*. London: Methuen, 1980.

Fowler, C.E. *The Coffer-Dam Process for Piers: Practical Examples from Actual Work*. New York: J. Wiley & New Sons, 1899.

Frazer, W. *A History of English Public Health 1834–1939*. London: Ballière, Tindall & Cox, 1950.

Gavin, H. *Sanitary Ramblings*. London, 1848; London: Frank Cass & Co., 1971.

Halliday, S. *The Great Stink of London*. Stroud: Sutton Publishing, 1999.

Humphreys, N., ed. *Vital Statistics: A Memorial Volume of Selections from the Reports and Writings of William Farr*. London: Sanitary Institute, 1885.

Jephson, H, *The Sanitary Evolution of London*. London: Fisher Unwin, 1907.

Martin, J.K.L. *Objections to the Tunnel Sewer, Proposed by the Metropolitan Sewage Manure Company, by Their New Bill, Together with an Alternative Plan, Consistent with the Recommendations of the Select Committee of the House of Commons*. London, 1847.

Mayhew, H. *London Labour and the London Poor: A Cyclopaedia of the Condition and Earnings of Those That Will Work, Those That Cannot Work, and Those That Will Not Work*. 4 vols., London, 1861–62.

Midwinter, E.C. *Victorian Social Reform*. London: Longmans, 1968.

Morris, R.J. *Cholera 1832: The Social Response to an Epidemic*. London: Croom Helm, 1976.

Nightingale, F. *Notes on Nursing*. London, 1860.

New Commission of Sewers: With Comments on the Reports and Evidence of the Sanitary Commission, by a Farmer, a Lawyer and an Ex-Commissioner. London, 1848.

Pike, R.E. *Human Documents of the Industrial Revolution in Britain*. London: George Allen & Unwin, 1966.

Pinks, W.J. *The History of Clerkenwell*. London, 1865.

Quennell, P., ed. *Mayhew's Characters*. London: Spring Books, 1967.

Razzell, P.E., and Wainwright, R.W., eds. *The Victorian Working Class: Selections from Letters to the "Morning Chronicle."* London: Cass, 1973.

Redfern, P. *Report to the Directors of the London Spring Water Company on the Organic Matters Found by Microscopical Examination of Waters Supplied from the Thames and Other Sources*. London, 1852.

Report on the Sanitary Conditions of the Labouring Population of Great Britain, by Edwin Chadwick. London: Poor Law Commission, 1842; Edinburgh University Press edition, 1965.

Robson, W.A. *The Government and Misgovernment of London*. London: Allen &
Unwin, 1939.

Sandström, G.H.E. *The History of Tunnelling*. London: Barrie & Rockcliff, 1963.

Simms, F.W. *Practical Tunnelling*, 3rd ed. London, 1877.

Simon, J. *Report on the Cholera Epidemic of 1854, as It Prevailed in the City of
London*. London, 1854.

————. *Report on the Last Two Cholera Epidemics of London as Affected by the
Consumption of Impure Water*. London, 1856.

Snow, J. *On the Mode of Communication of Cholera London*. London, 1849.

————. *On the Mode of Communication of Cholera London*, 2nd ed. London,
1855.

Tames, R.L.A., ed. *Documents of the Industrial Revolution*. London: Hutchinson
Educational, 1971.

Thwaites, J. *A Sketch of the History and Prospects of the Metropolitan Drainage
Question*. Southwark, 1855.

Trench, R., and Hillman, E. *London Under London*. London: Murray, 1993.

Wall, A.J. *Asiatic Cholera*. London: H.K. Lewis, 1893.

Wilkinson, J.J.G. *War, Cholera and the Ministry of Health: An Appeal to Sir Benjamin Hall and the British People*. London, 1854.

Wright, L. *Clean and Decent: The History of the British Bathroom and W.C.* London: G.J. Parris, 1957.

THE TRANSCONTINENTAL RAILROAD

—

Primary Sources

Bancroft Library—University of California, Berkeley (Crocker MS; Collis B.
Huntington MS; Mrs. Theodore Judah MS)

Department of Public Relations, Southern Pacific Company (Erle Heath MS)

Free Public Library, Council Bluffs, Iowa (Dodge MS and *Autobiography*)

Hopkins Transportation Library, Stanford University (Montague-Ives correspondence)

New York Public Library, Manuscript Division (W.H. Jackson, *Diary*, July 1866)

Union Pacific Historical Museum, Omaha, Nebraska (Reed MS)

University of Wyoming, Laramie, Wyoming (Casement MS)

Wyoming State Archives and Historical Dept., Cheyenne, Wyoming (Hezekiah Bissell MS)

Journals, Newspapers, and Periodicals

American Society of Civil Engineers, Transactions

Brown Quarterly

Congressional Globe

Council Bluffs Bugle

Daily American Flag (San Francisco)

Los Angeles Times

New York Times

New York Tribune

North American Review

Overland Monthly

Sacramento Union

San Francisco Bulletin

The *Times*

Union Pacific Magazine

Secondary Sources

Ambrose, S.B. *Nothing Like It in the World: The Men Who Built the Transcontinental Railroad*. New York, London: Simon & Schuster, 2000.

Ames, C.E. *Pioneering the Union Pacific*. New York: Meredith Corporation, 1969.

Bailey, W.F. *The Story of the First Transcontinental Railroad and Its Projectors, Construction and History*. Pittsburgh: Pittsburgh Printing Co., 1906.

Bain, D.H. *Empire Express: Building the First Transcontinental Railroad*, New York: Putnam, 1999.

Beadle, J.H. *The Undeveloped West*. Philadelphia: National Publishing Co., 1873.

Bowles, S. *Across the Continent: A Summer's Journey to the Rocky Mountains, the Mormons, and the Pacific States with Speaker Colfax*. Springfield, MA: S. Bowles & Co.; New York: Hurd & Houghton, 1865.

Cushing, R., and Moreau, J. *A Pictorial History of the Pacific Railroad*. Pasadena, CA: Pentrex, 1994.

Davis, J.P. *The Union Pacific Railway: A Study in Railway Politics, History and Economics*. Chicago: S.C. Griggs & Co., 1894.

Dodge, G.M. *How We Built the Union Pacific Railroad, and Other Railway Papers and Addresses*, Senate Documents, 61st Congress, 2nd Session, Vol. 59, 1791. Washington, D.C.: Government Printing Office, 1910.

———. *Personal Recollections*. Council Bluffs, IA: Monarch Printing Co., 1914.

Galloway, J.D. *The First Transcontinental Railroad: Central Pacific, Union Pacific*. New York: Arno Press, reprinted 1981.

Gibson, O. *The Chinese in America*. Cincinnati, OH, 1877.

Grinnell, G.B. *The Fighting Cheyennes*. New York: C. Scribner's Sons, 1915.

———. *Two Great Scouts and Their Pawnee Battalion: The Experiences of Frank J. North and Luther H. North in the Great West, 1856–1882, and Their Defense of the Building of the Union Pacific Railroad*. Cleveland, OH: A.H. Clark Co., 1928.

Griswold, W.S. *A Work of Giants: Building the First Transcontinental Railroad*. London: Frederick Muller, 1963.

Hogg, G.L. *Union Pacific*. London: Hutchinson, 1967.

Howard, R.W. *The Great Iron Trail: The Story of the First Transcontinental Railroad*. New York: G.P. Putnam's Sons, 1962.

Judah the Dreamer. Roseville, CA: Roseville Historical Society, 1989.

Klein, M. *Union Pacific: Birth of a Railroad, 1862–1893*. Garden City, NY: Doubleday, 1987.

Knight, O. *Following the Indian Wars: The Story of the Newspaper Correspondents Among the Indian Campaigners*. Norman, OK: University of Oklahoma Press, 1960.

MacCague, J. *Moguls and Iron Men: The Story of the First Transcontinental Railroad*. New York: Harper & Row, 1964.

MacClure, A.K. *Three Thousand Miles Through the Rocky Mountains*. Philadelphia: 1869.

Perkins, J.R. *Trails, Rails and War: The Life of General G.M. Dodge*. Indianapolis, IN: Bobbs-Merrill Co., 1929.

Poe, Co. O.M. *Report on Transcontinental Railways, 1883*. House Executive Document No.1, Part 2, 48th Congress, 1st Session, Washington, D.C.: Government Printing Office, 1884.

Report of the Commission and of the Minority Commissioner of the US Pacific Railway Commission, 8 vols., Washington, D.C.: Government Printing Office, 1887–88.

Report of the Joint Senate Committee to Investigate Chinese Immigration, Senate Report No. 689, 44th Congress, 2nd Session, Washington, D.C.: Government Printing Office, 1877.

Reports of the Government Directors of the Union Pacific Railroad Company, 1864–1884, Senate Executive Document No. 69, 49th Congress, 1st Session, Washington, D.C.: Government Printing Office, 1886.

Richardson, A.D. *Beyond the Mississippi: From the Great River to the Great Ocean: Life and Adventure on the Prairies*. Hartford, CT, 1867.

Rusling, J.F. *Across America*, New York: Sheldon & Co., 1874.

Sabin, E.L. *Building the Pacific Railway*. Philadelphia, London: J.B. Lippincott Co., 1918.

Seymour, S. *Incidents of a Trip Through the Great Platte Valley, to the Rocky Mountains and Laramie Plains, in 1866, with a Synoptical Statement of the Various Pacific Railroads*. New York: D. Van Nostrand Co., 1867.

Solomon, B. *Union Pacific Railroad*. Osceola, WI: MBI Publishing, 2000.

Stanley, H.M. *My Early Travels and Adventures in America and Asia*, 2 vols. London: Sampson Low and Marston, 1895.

Williams, J.H. *A Great and Shining Road*. New York: Times Books, 1988.

THE BROOKLYN BRIDGE

—

Primary Sources

Library of Rensselaer Polytechnic Institute, Troy, New York (Roebling Collection)

Special Collections of the Library of Rutgers University, New Jersey (Roebling Collection)

Journals, Newspapers, and Periodicals

American Artisan

American Railroad Journal

American Society of Civil Engineers, Transactions

Brooklyn Eagle

Chicago Times

Engineering (London)

Engineering News

Frank Leslie's Illustrated News

Harper's Franklin Square Library

Harper's New Monthly Magazine

Harper's Weekly

Illustrated London News

Institution of Civil Engineers, Proceedings

The *Iron Age*

Journal of the Franklin Institute

Mechanics (New York)

New York Evening Post

New York Herald

New York Journal of Commerce

New York Star

New York Sun

New York Times

New York Tribune

New York World

Scientific American

Scientific American Supplement

Trenton Daily State Gazette

Van Nostrand's Eclectic Engineering Magazine

Secondary Sources

Annual Reports of the Chief Engineer and General Superintendent of the East River Bridge, June 12, 1870. Brooklyn, New York: Eagle Book and Job Printing Department, 1870.

The Brooklyn Bridge 75th Anniversary Exhibition. New York: Brooklyn Institute of New York Arts and Sciences, Museums, 1958.

Communication from the Chief Engineer, W.A. Roebling, in Regard to the Method of Steam Transit over the East River Bridge, Made to the Board of Trustees, March 4, 1876. Brooklyn, NY: 1878.

Farrington, E.F. *Concise Description of the East River Bridge, with Full Details of the Construction.* New York: C.D. Wynkoop, 1881.

———. *History of the Building of the Great Bridge.* New York: Mooney, 1881.

Green, S.W. *A Complete History of the New York and Brooklyn Bridge.* New York: S.W. Green's Son, 1883.

Harris, F. *My Life and Loves.* Paris: published privately, 1922.

Hewitt, A.S. *Address on the Opening of the New York and Brooklyn Bridge, May 24, 1883.* New York: J. Polhemus, 1883.

Hildenbrand, W. *Cable-making for Suspension Bridges: With Special Reference to the Cables of the East River Bridge.* New York, D. Van Nostrand, 1877.

McCullough, D. *Brooklyn . . . and How It Got That Way.* New York: Dial Press, 1983.

———. *The Great Bridge.* New York: Simon & Schuster, 1982.

New York and Brooklyn Bridge, Proceedings 1867–1884. Brooklyn, NY: 1885.

Pneumatic Tower Foundations of the East River Bridge. Brooklyn, NY: Eagle Book and Job Printing Department, 1872.

Report of the Board of Consulting Engineers to the Directors of the New York Bridge Company. Brooklyn, NY: Standard Press Print, 1869.

Report of the Chief Engineer and General Superintendent of the New York and Brooklyn Bridge, June 1, 1884. Brooklyn, NY: Eagle Book and Job Printing Department, 1884.

Report of the Chief Engineer and General Superintendent to the Board of Directors of the New York Bridge Company, June 5, 1871. Brooklyn, NY: Eagle Book and Job Printing Department, 1871.

Report of the Chief Engineer of East River Bridge on Prices of Materials and Estimated Cost of the Structure, June 28, 1872. Brooklyn, NY: Eagle Book and Job Printing Department, 1872.

Report of the Chief Engineer of the New York and Brooklyn Bridge, January 1, 1877. Brooklyn, NY: Eagle Book and Job Printing Department, 1877.

Report of the Chief Engineer on the Strength of the Cables and Suspended Superstructures of the Bridge, Made to the Board of Trustees, January 9, 1882. Brooklyn, NY: Eagle Book and Job Printing Department, 1882.

Report of the Executive Committee, Chief Engineer and General Superintendent of the New York Bridge Company, June 1, 1872. Brooklyn, NY: Eagle Book and Job Printing Department, 1872.

Report of the Executive Committee, Chief Engineer and Treasurer of the New York Bridge Committee. Brooklyn, NY: Eagle Book and Job Printing Department, 1874.

Report of the General Superintendent of the New York Bridge Company. Brooklyn, NY: Eagle Book and Job Printing Department, 1873.

Report of the Officers of the New York Bridge Company to the Board of Directors, Brooklyn, NY: Eagle Book and Job Printing Department, 1875.

Roebling, J.A. *Report of John A. Roebling to the Presidents and Directors of the New York Bridge Company on the Proposed East River Bridge*. Brooklyn, NY: Daily Eagle Print, 1867.

————. *Report of John A. Roebling to the President of the New York Bridge Company on the Proposed East River Bridge*. Brooklyn, NY: Eagle Book and Job Printing Department, 1870.

Roebling, J.A. Sons Co., *An Illustrated Description of the New York and Brooklyn Bridge: Built Under the Direction of W.A. Roebling, etc.* Providence, RI: Livermore & Knight Co., 1899.

Roebling, W.A. *Specifications for Certain Steelwork Required for the Suspended Superstructure of the East River Bridge.* Brooklyn, NY: Eagle Book and Job Printing Department, 1881.

Schulyer, H. *The Roeblings: A Century of Engineers, Bridge Builders, and Industrialists.* Princeton, NJ: Princeton University Press, 1931.

Shapiro, M.J. *The Picture History of the Brooklyn Bridge.* New York: Dover Publications, 1983.

Smith, A.H., M.D., *The Effects of High Atmospheric Pressure, Including the Caisson Disease.* Brooklyn, NY: Eagle Book and Job Printing Department, 1873.

Steinman, D.B. *The Builders of the Bridge: The Story of John Roebling and His Son*, 2nd ed. New York: Harcourt, Brace & Co., 1950.

Stewart, E.C., ed. *Guide to the Roebling Collections at Rensselaer Polytechnic Institute and Rutgers University.* New York: Rensselaer Polytechnic Institute, 1983.

Trachtenberg, A. *Brooklyn Bridge: Fact and Symbol.* New York: Oxford University Press, 1965.

Weigold, M.E. *Silent Builder: Emily Warren Roebling and the Brooklyn Bridge.* New York: Associated Faculty Press Inc., 1984.

Young, J.H. *Lyrics.* New York: Funk & Wagnalls, 1889.

Younger, W.L. *Old Brooklyn in Early Photographs, 1865–1929.* New York: Dover Publications, 1978.

THE PANAMA CANAL

Primary Sources

Canal Zone Library-Museum (Special collection, deposited papers relating to construction)

Isthmian Historical Society, Balboa Heights, Canal Zone (Reminiscences of life and work during the construction of the Panama Canal, letters in response to a competition in 1963)

Library of Congress, Washington, D.C. (Philippe Bunau-Varilla papers; George Goethals papers; William Gorgas papers; Theodore Roosevelt papers; John F. Stevens papers; William H. Taft papers)

Journals, Newspapers, and Periodicals

American Society of Civil Engineers, Transactions
Engineering
Engineering Magazine
Engineering News
Evening Post
Harper's Weekly
Illustrated London News
Journal of the Franklin Institute
Longman's Magazine
Medical Record
Military Surgeon
National Geographic Magazine
New York Times
Pacific Historical Review
The *Times*

Secondary Sources

Abbot, H.L. *Problems of the Panama Canal, Including Climatology of the Isthmus, Physics and Hydraulics of the River Chagres, Cut at the Continental Divide, and Discussion of Plans for the Waterway*. New York: Macmillan Co., 1905.

Addresses at the de Lesseps Banquet Given at Delmonico's 1 March 1880. New York, 1880.

Albrecht, F.K. *Modern Management in Clinical Medicine*. London: Baillière, Tindall & Cox, 1946.

Anguizola, G. *Philippe Bunau-Varilla: The Man Behind the Panama Canal.*
Chicago: Nelson-Hall, 1980.

Beatty, C. *Ferdinand de Lesseps: A Biographical Study.* London: Eyre & Spottis-
woode, 1956.

Bennett, I.E. *History of the Panama Canal: Its Construction and Builders.* Washington, D.C.: Historical Publishing Co., 1915.

Bishop, F., and Bishop, J.B. *Goethals: Genius of the Panama Canal.* New York:
Harper & Bros., 1930.

Bunau-Varilla, P. *From Panama to Verdun: My Fight for France.* Philadelphia:
Dorrance & Co., 1940.

———. *The Great Adventure of Panama, Wherein Are Exposed Its Relation to the
Great War and the German Conspiracies Against France and the US.* London;
Garden City, N.Y.: Curtis Brown, 1920.

———. *The Nineteen Key Documents of the Panama Drama.* Paris, 1938.

———. *Panama: The creation, destruction and resurrection.* London: Constable,
1913.

Chamberlain, W.P. *Twenty-five Years of American Medical Activity on the Isthmus
of Panama, 1904–1929: A Triumph of Preventative Medicine.* Mount Hope:
Canal Zone, 1929.

Franck, H.A. *Things as They Are in Panama.* London: T. Fisher Unwin, 1913.

Gibson, J.M. *Physician to the World: The Life of General William C. Gorgas.*
Durham, N.C.: Duke University Press, 1950.

Goethals, G.W. *The Panama Canal: An Engineering Treatise.* New York:
McGraw-Hill Book Co., 1916.

Gorgas, M.D., and Hendrick, B.J. *William Crawford Gorgas—His Life and Work.*
London: William Heinemann, 1924.

Gorgas, W.C. *Population and Deaths from Various Diseases in the City of Panama,
by Months and Years, from November 1883 to August 1906.* Washington, D.C.:
Government Printing Office, 1906.

House Misc. Doc. 599, 50th Congress, 1st Session, Washington, D.C.: Government Printing Office, 1889.

———. *Sanitation in Panama.* New York and London: D. Appleton & Co., 1915.

Isthmian Canal Commission, *Annual Reports*. Washington, D.C.: Government Printing Office, 1904–1914.

Keller, U., ed. *The Building of the Panama Canal in Historic Photographs*. New York: Dover Publications, 1983.

Kimball, Lieut. W.W. *Special Intelligence Report on the Progress and Work on the Panama Canal During the Year 1885*. Washington, D.C.: Government Printing Office, 1886.

Lesseps, F.M. de, Viscount. *The Isthmus of Panama Inter-Oceanic Canal: M. le Comte de Lesseps at Liverpool, with the Address of the County, and the Speech of Captain Peacock, FRGS as to the Feasibility of Uniting the Atlantic with the Pacific Ocean*. Exeter: William Pollard, 1880.

———. trans. Pitman, C.B. *Recollections of Forty Years*, 2 vols. London: Chapman & Hall, 1887.

Manual of Information Concerning Employments for Service on the Isthmus of Panama. Washington, D.C.: Government Printing Office, 1909.

McCullough, D. *The Path Between the Seas: The Creation of the Panama Canal*. New York: Simon and Schuster, 1977.

Munchow, Mrs. E.U. von. *The American Woman on the Panama Canal, 1904–1916*. Panama: Star & Herald Co., 1916.

Nelson, W. *Five Years at Panama: The Trans-isthmian Canal*. London, New York: Sampson Low & Co., 1891.

Papers Relating to the Panama Canal. Washington, D.C.: Government Printing Office, 1909.

Report of the Board of Consulting Engineers for the Panama Canal. Washington, D.C.: Government Printing Office, 1906.

Report of the Isthmian Canal Commission, 1889–1901, Senate Document 222, 58th Congress, 2nd Session, Washington, D.C.: Government Printing Office, 1904.

Rodrigues, J.C. *The Panama Canal: Its History, Its Political Aspects, and Financial Difficulties*. New York: Charles Scribner's Sons, 1885.

Rogers, Lieut. C.C. *Intelligence Report of the Panama Canal, March 30, 1887*.

Roosevelt, F.D. *Collections: The Public Papers and Addresses of Franklin D. Roosevelt*, 13 vols. New York: Random House, 1938–50.

Schonfield, H.J. *Ferdinand de Lesseps*. London: Herbert Joseph, 1937.

Sibert, W.L., and Stevens, J.F. *The Construction of the Panama Canal*. New York and London: D. Appleton & Co., 1915.

Smith, G.B. *The Life and Enterprises of Ferdinand de Lesseps*. London, W.H. Allen & Co., 1893.

Stevens, J.F. *An Engineer's Recollections*. New York: McGraw-Hill, 1936.

Subject Catalogue of the Special Panama Collection of the Canal Zone Library-Museum. The History of the Isthmus of Panama as It Applies to Interoceanic Transportation. Boston: G.K. Hall & Co., 1964.

Underwood, W.L. *The Mosquito Nuisance and How to Deal with It*. Boston, 1903.

Weir, H.C. *The Conquest of the Isthmus*. New York & London: G.P. Putnam's Sons, 1909.

THE HOOVER DAM

—

Primary Sources

Bancroft Library, University of California, Berkeley (Henry J. Kaiser Collection; Six Companies Corporate Records)

Boulder City Library, Boulder City, Nevada (Hoover Dam and Boulder City Collection)

Bureau of Reclamation Lower Colorado Region Library, Photo Laboratory and Map Collection, Boulder City, Nevada

Dennis McBride, Boulder City, Nevada (Boulder City Oral History Project Collection)

Federal Records Center and National Archives, Denver, Colorado (Annual Project Histories, Boulder Canyon Project)

James R. Dickenson Library, University of Nevada–Las Vegas, Las Vegas, Nevada (Special Collections—Ragnald Fyhen deposit)

Journals, Newspapers, and Periodicals

American Society of Civil Engineers, Transactions
Business Week
Civil Engineering
Compressed Air Magazine
Electric Engineering
Engineering
Harper's Magazine
Industrial Worker
Institution of Civil Engineers, Proceedings
Las Vegas Age
Las Vegas Evening Review-Journal
Los Angeles Examiner
Military Engineer
National Geographic Magazine
New Reclamation Era
New York Times
Reclamation Era
San Francisco Examiner

Secondary Sources

Allen, M.V. *Hoover Dam and Boulder City.* Shingletown, CA: privately printed, 1983.

"The Dam," *Fortune* 8 (September 1933), 74–88.

"The Earth Movers I," *Fortune* 28 (August 1943), 99–107, 210–14.

"The Earth Movers II," *Fortune* 28 (September 1943), 119–22, 219–26.

"The Earth Movers III," *Fortune* 28 (October 1943), 139–44, 193–9.

Gates, W.H. *Hoover Dam, Including the Story of the Turbulent Colorado River.* Los Angeles: Wetzel Publishing Co., 1932.

Grey, Z. *Boulder Dam*. London: Transworld Publishers, 1967.

The Hoover Dam Documents. Washington, D.C.: Government Printing Office, 1948.

Hoover, H. *The Memoirs of Herbert Hoover*, 3 vols. New York: Macmillan, 1952.

Kant, Candace. *Zane Grey's Arizona*. Flagstaff, AZ: Northland Press, 1984.

McBride, D. *In the Beginning: A History of Boulder City*. Boulder City, NV: Boulder City Chamber of Commerce, 1981.

Moeller, B.B. *Phil Swing and Hoover Dam*, Berkeley, CA: University of California Press, 1971.

Pettitt, George Albert. *So Hoover Dam Was Built*. Berkeley, CA: Lederer, Street & Zeuss, 1935.

Stevens, J.E. *Hoover Dam: An American Adventure*. Norman, OK, and London: University of Oklahoma Press, 1988.

The Story of Boulder Dam. Washington, D.C.: Government Printing Office, 1941.

Thomson, F. *The I.W.W. Its First 50 Years, 1905–1955: The History of an Effort to Organize the Working Class*. Chicago: Industrial Workers of the World, 1955.

Wilson, N.C., and Taylor, F.J. *The Earth Changers*, Garden City, NY: Doubleday, 1986.

Wollet, W. *Hoover Dam: Drawings, Etchings and Lithographs*. Santa Monica, CA: Hennessey & Ingalls, 1986.

ILLUSTRATION CREDITS

Page i: Getty Images/Hulton Archive

Page iv: Courtesy of Boulder City Museum and Historical Association
Getty Images/Hulton Archive

Page v: Getty Images/Hulton Archive

Page 42: Illustrations by Sarah Underwood

Page 154: © 2003 by Shirley Burman

Page 194: © 2003 by Mapgrafix; original map supplied courtesy of
Panama Canal Authority

PLATE SECTION I

1. Getty Images/Hulton Archive
2. Special Collections, University of Bristol
   Getty Images/Hulton Archive
3. © Brunel University
   Getty Images/Hulton Archive
4. Getty Images/Hulton Archive
   Getty Images/Hulton Archive
5. Crown copyright: RCAHMS
   Getty Images/Hulton Archive
6. New York Historical Society/Bridgeman Art Library
   Special Collections and University Archives, Rutgers University
   Libraries
   Special Collections and University Archives, Rutgers University
   Libraries
7. Special Collections and University Archives, Rutgers University
   Libraries
   The Stapleton Collection/Bridgeman Art Library
   © Historical Picture Archive/CORBIS
8. © Charles E. Rotkin/CORBIS
   Getty Images/Hulton Archive

PLATE SECTION 2

—

1. Getty Images/Hulton Archive
   Getty Images/Hulton Archive
   Pinwell, George John, Death's Dispensary; © 1985 by Philadelphia
   Museum of Art: The William H. Helfand Collection
2. Getty Images/Hulton Archive
   Getty Images/Hulton Archive
   Getty Images/Hulton Archive
3. (Clockwise from top left) Getty Images/Hulton Archive; Getty
   Images/Hulton Archive; Collection of Jeffrey Moreau; Collection of
   Jeffrey Moreau
4. Collection of Jeffrey Moreau
   Getty Images/Hulton Archive
   Getty Images/Hulton Archive
5. Courtesy of Panama Canal Company
   Getty Images/Hulton Archive
   Courtesy of Panama Canal Company
6. Getty Images/Hulton Archive
   Getty Images/Hulton Archive
7. Courtesy of Boulder City Museum and Historical Association
   Courtesy of Boulder City Museum and Historical Association
8. Courtesy of Boulder City Museum and Historical Association
   Courtesy of Boulder City Museum and Historical Association

ACKNOWLEDGMENTS

I AM INDEBTED to many specialists who made the television series and book *Seven Wonders of the Industrial World* possible. Numerous historians and scientific experts advised the BBC team over the years and it would not be possible to thank them all. I also owe a great deal to the studies of the many scholars cited in the references and to the team of archivists, led by Nick Barratt, without whose detailed investigation into many primary sources and manuscripts this research could not have been completed. The BBC team received wonderful support from a large number of specialist archives in Britain and America, and these are credited in the Bibliography.

In completing this manuscript I would particularly like to thank the following experts for kindly giving valued time to read and comment on different chapters: Professor Angus Buchanan, Centre for the History of Technology, University of Bath; Dr. Stephen Halliday of the Buckinghamshire Business School; Dennis McBride, Curator at the Boulder City Museum and Historical Association; Rolando Cochez at the Panama Canal Authority; Professor Roland Paxton at the Heriot-Watt University; and Professor Robert McWilliam, Senior Curator Engineering at the Science Museum, London.

At the BBC, I am grateful to the large team of talented producers for many enjoyable discussions about the story lines and the characters while making the drama documentary series. Executive Producer Jill Fullerton Smith brought the project to life and has been a great partner and friend throughout, along with producers Christopher Spencer, Paul Wilmhurst, Paul Bryers, Ed Bazalgette, Philip Smith, and Mark Everest—all of whom contributed to the ambition and momentum of the series. As Creative Director, Science, John Lynch was a valued supporter whose encouragement was much appreciated. Research and analysis could not have been completed without excellent input from the following assistant producers: Simon Winchcombe, Alison Draper, Sara Cropley, Lucy Bowden, Amy Leader, and Zoe Elliot. Finally, this book could not have been written without the generous assistance of Debbie Holder, Francesca Franklyn, and Rachel Wright.

At Fourth Estate, it was a pleasure, as always, to work with Christopher Potter, Nicky Eaton, and Courtney Hoddell. Thank you, too, to Nick Davies for his skilled editorial judgment and insight at every stage in the production of the manuscript and for keeping me on track through the tight deadlines. Thank you, too, to Sophie King for her excellent picture research. At Curtis Brown, Giles Gordon provided much valued advice on the project over many months.

Finally, thank you to Martin Surr, for keeping the world turning while I was writing, and to Julia Lilley, whose generous support and great faith in the project made this book possible.

INDEX

Page numbers in *italics* refer to illustrations.

INDEX

Page numbers in *italics* refer to illustrations.

Dreams of Iron and Steel

by Deborah Cadbury

About the author

About the book

Read on

Insights,
Interviews
& More . . .

Profile of
Deborah Cadbury

DEBORAH CADBURY grew up in a house where creativity and self-expression took center stage. "It was a very happy home," she remembers, "with my father busy in his garden and my mother painting." It became clear early on that Cadbury's own interest was in stories. She was fascinated by story, plot, and character—it was her medium. "If I got a book and I loved it, I wouldn't just read it, I'd devour it. I loved analyzing how the stories were put together."

However, she was advised at school to take advanced placement in science, something she later regretted. Her scientific bent, combined with an interest in people and what makes them behave the way they do, led her to study psychology at the University of Sussex. "When I got there, I found myself taking Experimental Psychology, where the whole thing seemed reduced to a study of rats and electric shocks. I was horrified that anyone thought this would have anything to do with human behavior." In the end, she was allowed more or less to devise her own course, studying religion and philosophy, before taking postgraduate studies in anthropology at Oxford.

Finally when it came to a career choice, she discovered the BBC ran a research trainee scheme. It seemed like an ideal first step although she was disheartened to realize there were thousands of applicants. On the day of the interview, she grabbed a little red velvet beret as she left the house. Years later, one of the people who had been on the panel said that over three days of interviewing, the one thing they could agree on was "the one with

❝ A group of us were very interested in turning science into storytelling. ❞

the hat." "I always felt that the hat got the job," admits Cadbury today.

She spent two varied years working on current affairs, arts, and science programs, learning how to go out shooting, direct film, and develop story. "This led to producing on [the British TV show] *Horizon*, where a group of us were very interested in turning science into storytelling. Instead of typical dry science essays, we realized you could make the programs character-driven, engaging the audience in how the scientists had their ideas and the steps they took to realize them." The program that gave birth to her writing career was her Emmy-winning *Assault on the Male*. "It spurred an international scientific research effort as people saw there was a serious problem with chemicals that mimic human hormones." Cadbury's related book, *The Feminization of Nature*, explored these issues in greater depth than was possible in a TV program.

The solitariness of writing is a dramatic contrast with the teamwork of TV producing. To enjoy doing both, Cadbury claims to be something of a chameleon. Her second book, *Terrible Lizard*, happened by chance. While reading a book about dinosaurs to her six-year-old son, she came across a footnote saying that Gideon Mantell, one of the first British dinosaur hunters, had kept a diary that had never been published. "The existence of such a thing was enormously exciting. I took a day off work to track it down in a small private library in Lewes. The archivist pulled the diaries from the back of a cupboard, covered in dust and ▶

JERRY BAUER

SNAPSHOT

BORN
....
August 1955 in London

EDUCATED
....
Guildford Grammar School; University of Sussex; University of Oxford

CAREER
....
BBC research trainee 1978–79; researcher 1979–82; TV producer 1983 onward; author

PREVIOUS WORKS
....
The Feminization of Nature, 1997

The Dinosaur Hunters, 2000

The Lost King of France, 2003

3

Profiles *(continued)*

◀ tied in ribbon. I opened them and was immediately transported to Mantell's Georgian world." Spurred to investigate further, her own book and a related TV series was born.

The Lost King of France sprang from her awareness of the DNA testing of the various pretenders who claimed to be the son of Marie-Antoinette. The turning point in her subsequent research came with the discovery of the journal written by the boy's sister, Marie-Thérèse. "I was so stirred by it that I felt compelled to bring the truth to light. In a novel, an editor would dismiss the events out of hand. But it was all true. It was very powerful to write."

As she intertwined her TV career with writing, a new and satisfying challenge came with a series on the Industrial Age. "It was a time of radical change when the environment was altered on an epic scale, a time of optimism and a sense that man could do anything. Great canals were built to join oceans, railways were built across continents, bridges were spanning greater distances than ever before. There was a sense that man was controlling the environment. That was really what was so inspiring about the series." She wrote the book *Dreams of Iron and Steel* while producing the TV series. Cadbury was able to use the book to get across the depth and the detail that inevitably had to be omitted from each 50-minute drama. Here were the characters that made history and the dramas and obstacles that beset them. Rich material for a writer . ∽

Deborah Cadbury's
Top Ten Favorite Books

1. **Pride and Prejudice**
   Jane Austen

2. **Citizens**
   Simon Schama

3. **Doctor Zhivago**
   Boris Pasternak

4. **Alias Grace**
   Margaret Atwood

5. **Justine**
   Lawrence Durrell

6. **The Sea, the Sea**
   Iris Murdoch

7. **Schindler's List**
   Thomas Keneally

8. **The Chronicles of Narnia**
   C. S. Lewis

9. **Darwin**
   Adrian Desmond and James Moore

10. **Birdsong**
    Sebastian Faulks

A Note from
Deborah Cadbury

Dreams of Iron and Steel tells the dramatic
stories of great achievement that made the
modern world. Each chapter highlights the
role of one of the leading pioneers of the nineteenth and early twentieth centuries who were
prepared to sacrifice everything for their
dream. For me, this gives the stories their great
appeal. The central themes are universal;
anyone who has ever struggled to realize his or
her ambitions can relate to these epic feats as
the legendary pioneers of the nineteenth
century take on the impossible.

The scale of their vision is breathtaking.
This was an era when the timeless traditions of
centuries were swept away and the landscape
was being changed on a spectacular scale as
the world itself became a smaller place. The
Panama Canal cut a swathe across the Isthmus
of Panama and joined the Atlantic and Pacific
oceans for the first time—shrinking the entire
globe since it was no longer necessary to face
the dangerous seas around the Horn. The
Transcontinental Railway, the first railway
to stretch right across a continent, united
America from coast to coast. Isambard
Kingdom Brunel's "Great Ship," also known
as the "Crystal Palace of the Seas," would
join the two ends of the British empire.

At the heart of each story were the practical
visionaries who rose to become some of the
most celebrated men of their generation. I
delved into American and British archives to
retrace their steps and their thinking and to
uncover the true stories behind the scenes as
they created a "wonder" of the modern world.
The more I found out, the more drawn I

became to the key characters who were quite prepared to drive themselves on, never losing faith in their vision—whatever the price to be paid. For instance, take the Roeblings, who designed the longest suspension bridge in the world at Brooklyn: first the father, John Roebling, and then his son, Washington, were to make the ultimate sacrifice. Brunel staked everything, driving himself almost to self-destruction as he struggled to bring his Great Ship to life. Ferdinand de Lesseps, the hero of Suez, a man so fêted in French society he was known as *Le Grand Français*, was ruined by the Panama affair; his son too paid a high price for his loyalty.

The archives also provide revealing details about the unsung heroes, the workers who made the wonders possible. With labor so cheap, there were gruesome accidents and horrific working practices. The workers' achievements were all the more remarkable given the technology that was available at the time. To build the Bell Rock Lighthouse out in wild northern seas off the east coast of Scotland at the beginning of the nineteenth century, Robert Stevenson and his men had little more than their own hands and simple pickaxes to work with. To build the double hull of the Great Ship, Brunel employed children, some as young as nine or ten, to work with hot rivets between the narrow spaces of the hull.

Yet the results of all this labor were awe-inspiring—and nowhere more so than in America. Take railways: by 1850 there were 9,000 miles of track—just ten years later, over 30,000 miles. New York took on its modern form as the Brooklyn Bridge spanned the East River in one giant leap. The whole of the southwestern desert was able to flower as Hoover Dam towered over the Colorado River and provided the energy to light up California. Whatever the difficulties, man found a way to win through and conquer vast oceans and wild rivers, and open up the continents. For me, this made the stories very inspiring.

—*Deborah Cadbury*

The Making of the
TV Series *Seven Wonders of the Industrial World*

CHOOSING THE Seven Wonders was a difficult task. It became clear that they should be determined by a combination of which was the most important breakthrough in terms of technical advances for its day and which story had the most dramatic potential with a compelling narrative arc for the key characters. "The ones we chose really stood out, given the criteria we'd chosen to apply," Cadbury remembers. She and Executive Producer Jill Fullerton-Smith then faced a number of hurdles before the programs could be made. "We were very apprehensive because we only had a modest documentary budget and neither of us had made a drama so it was a bit of a leap."

Their first task was to devise a format. "We knew we wanted to tell the stories as drama because it would be vivid and immediate and allow us to immerse the viewers in a different world. We wanted to highlight the pioneers' vision and the lengths they would go to to achieve it. At the same time, we wanted authenticity." Their challenge was to find a way of devising scripts that would feel real and capture exactly what happened. They asked an archivist, Nick Barratt, to track down all the original minutes of meetings, letters, and other primary research material. "We then came at the scripts in a new way for drama. The directors blocked out all the real quotes that could be found in the archives and used this to produce the first draft." Their object was to create an observational documentary, taking the viewer back 150 years

> ❝ We were very apprehensive because we only had a modest documentary budget and neither of us had made a drama so it was a bit of a leap. ❞

to witness the pioneer himself and the troubles he faced. They had to work out which scenes a crew would have been eye-witness to, what kind of action they should cover. Making it feel convincingly like a documentary made at the time allowed them to play to the strength of the material coming through the archives.

Each Wonder then inevitably presented its individual problems when it came to making the script work as drama. "Panama didn't naturally lend itself to drama because it's a broken-backed story where your cast die halfway through and a new set of characters has to be brought in. With the Transcontinental, there's a huge cast of characters and no one pioneer driving them. There was very little archive material relating to Joseph Bazalgette, who was such a modest man he didn't seem to keep letters, a diary, anything. It was very hard to put his scenes together convincingly, working out what might have been in his mind and the problems he was up against."

The production of the various episodes threw up more practical problems. "Brunel's Great Ship did not exist and had to be created with the help of our very talented graphics designer, Gareth Edwards, who used archive photography and 2-D computer graphics. When Director Chris Spencer and his team went to film the lighthouse scenes, they needed stormy seas but got two weeks of water as calm as a millpond. Director Mark Everest and his production team went to Nevada to find the hottest summer on record since the Hoover Dam itself was built. Many of the dam workers had died of heatstroke, so we were very worried for the crew. They were working long days in difficult filming conditions." ▶

> " Each Wonder then inevitably presented its individual problems when it came to making the script work as drama. "

The Making of the TV Series *(continued)*

◄ Having been so closely involved with the series and writing the accompanying book, Cadbury found herself extremely moved when she saw the first cut of the films. "It felt so real after living with the characters for so long, the directors brought them to life brilliantly."

Out of all seven, does Cadbury have a favorite? "I love them all, but particularly the lighthouse. The pioneers' isolation in those wild northern seas and their determination to light up the ocean in this most dangerous way, really with just their bare hands, is a moving testament to what ordinary men can achieve." ～

❝ Out of all seven, does Cadbury have a favorite? 'I love them all, but particularly the lighthouse.' ❞

A Critical Eye

WHAT STRUCK REVIEWERS was Cadbury's extraordinary ability to re-create events and characters in the most vivid way. Writing in the *Guardian*, Manjit Kumar said, "While 'practical visionaries' such as Brunel, Stevenson and Roebling may have been 'taking risks and taking society with them as they cut a path to the future,' Cadbury never forgets those risking their lives just to survive. What makes this book a compelling read is the heroism and desperation of ordinary men." The TV series received nominations from the British Academy of Film and Television Arts, the Royal Television Society, and the Broadcast Awards. *The Times*'s Graham Stewart wrote of the TV version: "At last a style of drama documentary has been conceived that attains the artistry of great drama with the faithfulness demanded of an honest documentary. The acting is convincing and money has been spent ensuring sets and effects worthy of a mid-budget Hollywood film. Importantly, the dialogue given to the characters has been sourced to their recorded pronouncements. The only appropriate response as the credits roll at the end of *Seven Wonders of the Industrial World* is a burst of applause." Elsewhere it was seen as "a fine flavorsome plum"(Paul Hoggart, *The Times*), "a real return for your license fee" (*TV Times*), and simply "superb" (*Guardian*). ◐

The Seven Wonders
of the Ancient World

1. **The Great Pyramid of Giza**—the tomb of the Ancient Pharaoh Khufu close to the city of Memphis, Egypt

2. **The Hanging Gardens of Babylon**—a legendary palace with beautiful gardens built on the banks of the Euphrates River by King Nebuchadnezzar II

3. **The Statue of Zeus at Olympia**—a vast statue of the king of the gods by Pheidias

4. **The Temple of Artemis at Ephesus**—a temple built in honor of the Greek goddess of hunting

5. **The Mausoleum at Halicarnassus**—a unique tomb built for Maussollos, Persian satrap of Caria

6. **The Colossus of Rhodes**—a massive statue of Helios the sun god, built at Rhodes harbor

7. **The Lighthouse of Alexandria**—a lighthouse built by the Ptolemies of Egypt off the island of Pharos

Have You Read?

THE LOST KING OF FRANCE
by Deborah Cadbury

"Absolutely stupendous.... This is history as it should be. I can't praise it highly enough. It is stunningly written. I could not put it down. This is the best account of the French Revolution I have ever read."

Alison Weir, author of *Mary Queen of Scots*

"Unputdownable.... Deborah Cadbury succeeds in conveying the human tragedy of the French royal family's end more emotively than any other writer. Added to that, her book has the gripping pace of a thriller and offers the ultimate satisfaction in revealing how a historical mystery has been solved by modern science. For sheer escapism and some fascinating insights into history, I cannot recommend this too highly."

Maureen Waller,
author of *1700: Scenes from London Life*

"Outstanding... the action races forward with sumptuously judged pace equal to that of any top-rate thriller. Cadbury's masterstroke is to unfurl the Revolution's black dramas through the eyes of a small boy who is expelled from Versailles' carpet of flowers into a life of unrecognizable, frightening chaos.... Wrapped around the Dauphin's story comes a gripping account of the Revolution itself."

George Lucas, *Financial Times*

TERRIBLE LIZARD
by Deborah Cadbury

"This is a tale of intrigue and deception, of burning ambition and failed dreams. The ▶

Have You Read? *(continued)*

◀ bitter clashes between the men who dominated nineteenth-century geology are exquisitely portrayed by Deborah Cadbury in this scholarly yet exhilarating book."

Steve Connor, *Independent*

"Cadbury is a wonderful writer, weaving natural history, human history and science together in a smooth, flowing tapestry that keeps you turning the pages as if her book was a thriller, which of course it is, as we track the rise and fall of the dastardly Richard Owen, as unscrupulous a villain as ever appeared in Victorian fiction.... An important book."

Richard Ellis, *The Times*

"Deborah Cadbury has made this true story a compulsive read.... There is a wonderful character list straight out of Dickens, with toadying worthies, academic eccentrics and pompous priests, along with a heady mixture of ambition and skulduggery."

Douglas Palmer, *The Herald* (Scotland)